World Climate

T. F. Gaskell and Martin Morris

World Climate

the weather,
the environment
and man

with 53 illustrations

Thames and Hudson

© *1979 by T. F. Gaskell and Martin Morris*

All rights reserved. No part of this publication may be reproduced or transmitted in any form or by any means, electronic or mechanical, including photocopy, recording, or any information storage and retrieval system, without permission in writing from the publisher.

Library of Congress Catalog card number 79–66128

Printed and bound in Great Britain by Butler & Tanner Ltd, Frome and London

Contents

Preface

Everybody has his own view about climate, but few are well enough versed in the details of weather mechanisms, and the interaction of the sea, earth and atmosphere, to distinguish fact from fiction. When I was asked by Thames and Hudson to write a book for the layman that would elucidate some of the physical principles that determine hot and cold, rain and fine, I somewhat reluctantly agreed. In order to add some meteorological experience to my oceanographic and geophysical knowledge I persuaded Martin Morris, Head of the London Weather Centre, to assist with the work.

It would not have been possible to collect all the material together in a comparatively short time without reference to the two large volumes (1,500 pages or more) of *Climate: Present, Past and Future* by Professor H. H. Lamb. My thanks to Professor Lamb are unbounded, both for being allowed to use his great work and for the encouragement he gave when informed of the project. If any reader of this short account wishes to delve further into historical, archaeological, meteorological or geophysical detail, Lamb's volumes are not only a mine of eminently readable information, but also contain detailed references and many interesting tables of such things as past temperatures which show that the weather has been worse than now in the historical past. The measurements in the text have been given in those units which, it is hoped, are most familiar to the reader.

Thanks are also due to Dr W. W. Kellogg of the National Center for Atmospheric Research at Boulder, Colorado, John Mason and David Houghton of the British Meteorological Office at Bracknell, David Hibbert and Bill Ireland of the more commercial side with International Meteorological Consultant Services, and Professor Joe Goldman of Houston who have all supplied written and verbal advice. Thanks are also due to those librarians and officers of the United States Coast Guard, United States Geological Survey, the Scott Polar Research Institute and the Compagnie Française du Pétrole who kindly produced lists of books and papers.

As is usual in writing a book, a heavy load of reading pencil scrawl falls on those who make it legible with the typewriter, and all thanks to Siobhan Rice and Louise Walker for their efficient translation.

Climate and weather are always with us, and they not only control living habits but also much of our conversation, especially in places like the European and American temperate zones where variety is the order of the day. We may now be at a point in evolution when it is necessary to find out what really makes the atmosphere tick. The collaboration of oceanographers, meteorologists and geophysicists is needed, and it is to be hoped that politicians conferring on the law of the sea will not restrict oceanic research in their coastal zones, since climate is a world-wide problem which affects small nations as well as the large ones, developing countries and those well established economically. This is truly a subject which knows no boundaries, especially man-made ones.

T. F. G.

1 Facts and fluctuations

You can produce oil from the North Slope of Alaska or from the Arabian desert without freezing to death or suffering from heat stroke, but in spite of central heating and air-conditioning, mankind in the latter part of the twentieth century is still at the mercy of the weather and of variations of climate. For those who live in temperate regions, the weather is fickle: the driest summer for many years in Britain, with associated rationing of water, can be followed by a very wet winter and spring and a cold summer; in the USA a very dry season reminds farmers of the dust-bowl catastrophe of the 1930s. Even in stable tropical regions where the seasons are more clearly defined, there are continual worries about drought and the consequent shortage of food for an ever-increasing population. Whether it is the effect on the basic necessities of life or merely on the enjoyment of leisure activities, the weather and climate are of interest and importance to everybody.

It is not surprising that almost everyone has firm opinions about the weather and especially its changes and what causes them and the omens that precede them. A spell of cold weather will convince many people that a new ice-age is starting, and even rapid alternations in seasons of exceptional hot or cold or wet or dry weather will lead to forecasts of world-wide upsets in the climate in which we grew up.

Is there a present danger that a new ice-age is about to descend on the northern parts of America and Europe? Are there any ways of seeing the trends of some reliable climatic indicator in time to avoid catastrophe, or even, with our powerful scientific and engineering tools, to take some suitable avoiding action? The evidence of ice-ages in the last million years is incontrovertible, being written clearly in the geological record of easily examined rocks and the debris from ancient glaciers – to say nothing of deep-frozen mammoths in the Siberian tundra. Similar geological evidence shows that ice-ages have occurred and recurred over a period of several hundred million years, so that they are part of the climate pattern of the Earth. Although we do not know exactly what starts an ice-age, or alternatively what switches it off, we are sure that ice-ages are part of some recurring feature of the Earth's behaviour and that, with careful investigation, the mechanism of the ice-age phenomenon can be discovered.

While all scientific leads are being followed to discover what determines climate, it is possible that modern weather forecasting may provide some information to suggest when a new ice-age is approaching. Weather forecasts, as explained in the following chapters, are based on a world-wide network of observation combined with comparison of what has happened in the past. The trouble is that the past record shows that the weather pattern is not a simple progression from one situation to another. A hot summer may be followed by a cold winter or a mild one; droughts may be averaged out by heavy rain the next year, or they may continue for many seasons. Storm fronts may stop for days after approaching at constant speed, and they may suddenly confound the forecaster by irritating accelerations. It is as if the weather itself does not know what it is going to do next.

If the weather is so undecided, how can we predict the future? Some say that animals, less artificial in their communion with nature than is man, can tell when it is going to rain, and maybe there are still some extra senses that scientists have missed. Professor Blainey, in his book *Triumph of the Nomads*, tells of Australian aboriginals who believed that they were protected by benevolent ancestors who could hasten the coming of rain through the rituals and ceremonial blood sacrifices carried out by the tribal rainmaker.

Better forecasts were expected in the dark ages in Europe when the Twelfth Night 'bean-feast' turned up the bean king as the lucky man who was served the slice of cake containing a bean. The popular belief was that the weather for the ensuing twelve months was determined month by month by what was experienced on each of the twelve days following Christmas. The chief duty of the bean king may well have been the performance of magical ceremonies in order to provide good weather for the year. Probably part of his job was to explain why the prognostications were not exactly in line with what occurred. The forecaster's life was always a difficult one, especially since many local sages and self-professed experts have produced a whole mass of country lore, based on behaviour of animals and plants, the colour of the sky, feelings of rheumatic joints, old broken limbs and so on, with which to confound the professional meteorologist. He is under the great handicap of having to put his forecasts out regularly, and ahead of time, so that his errors cannot conveniently be forgotten as is so often the case with the errors of his critics. However, sometimes useful generalizations about the weather patterns lie behind the old wives' tales.

Considerable variations of rainfall and temperature occur from year to year and from one decade to the next, although the general pattern which emerges from the hundred years or so of actual recorded values, together with the various extrapolations that have been made into the past, is that the world's climate has remained much the same since the ending of the last ice-age some 10,000 years ago. There have been historical reports of famines produced by a run of bad harvests due to lack of rain, floods or other climatic catastrophes. Now that the world population is growing so quickly – largely because of great advances in medicine – the forecasting of weather will become more important, since even marginal changes in climate may have sufficient effect on agriculture to produce disaster. While uncertainties remain about the direction in which the world's climate is moving – if it *is* in fact changing appreciably – it is impossible to persuade governments to alter such factors as the pattern of food-growing to cope with the population explosion. Much greater certainty exists about the future occurrence of earthquakes in places such as Agadir or San Francisco, but no one in authority can be persuaded to order the evacuation of the cities. How then can authority be expected to take action on anything so uncertain as future climatic trends until a better understanding is achieved of what causes these trends, and how large must be fluctuations from normal before they are significant indicators of change?

One of the ways in which progress is being made is in studying the weather today. The normal climate depends upon the rotation of the Earth and its accompanying atmosphere and on the heat received from the sun. The circulation of the atmosphere at different heights provides the basis for smaller circulations of cyclones and anticyclones, which in turn bring with them evaporation and precipitation of water. Atmospheric circulation can be studied now better than ever before, because of the widely distributed meteorological stations measuring temperature, barometric pressure, rainfall, wind speed and direction. The fixed stations are backed by balloons carrying instruments into the upper reaches of the atmosphere and by a view from outside of cloud movements obtained by satellites. The exercise of making forecasts for a few hours, half a day, three days, six days and so on, and producing regular pictures of the air movement over the globe must be a step in the right direction to understanding weather mechanisms. Theoretically, if we knew the exact state of the Earth and its atmosphere at any one instant, and if we had a big enough computer, we could move ahead step by step for ever and say what the future weather will be, if we also knew all about the inside of the Earth and the sun.

The first step in monitoring weather and climate is to look at the Earth in its role as a planet and part of the solar system.

The sun is the great supplier of the energy that we use on Earth. We use wood and coal fires to cook our food and to warm ourselves, and oil to transport us around the world, but coal and oil are fossil fuels, produced from plant and animal life millions of years ago, and so they are just capital that has been saved by special natural processes of accumulation. It is comforting to note that the energy from the sun which falls every minute on the Earth is thousands of times the amount we use, although with a rapidly growing population and rising standards of living, which demand high energy consumption, some form of human plan is needed today. The limited coal and oil stocks must be used to move, first, to nuclear energy and then to solar energy in order to keep the human race going.

The heat from the sun has a continuous effect on the weather and it controls the climate. There are exchanges of heat between the sun and the atmosphere: some heat is absorbed by the land and radiated back again at night; the oceans act as a large heat store and give back heat to the air under certain conditions; the water in the atmosphere acts as a powerful and sometimes explosive heat exchange because of the heat produced when water condenses from vapour to liquid. Not all the sun's warmth which falls on the Earth is effective in providing heat.

The albedo (literally, 'whiteness' – the proportion of the solar radiation which is scattered or reflected from the Earth's surface) plays an important part in climatic history. Clouds reflect a good deal of the sun's rays, as also do snow and icefields; and there are wide variations of the albedo over land, desert areas absorbing much less energy than those parts which are covered with vegetation. Dust in the atmosphere tends to cut out the energy from the sun; the huge volume of dust shot into the atmosphere in 1883, when the volcano Krakatoa erupted with the force of a 50-megaton hydrogen bomb, remained in the atmosphere for several years, its presence being shown by the beautiful sunsets it caused.

The Earth changes its orbit round the sun slightly on account of the attraction of other planets which occupy changing relative positions in the course of hundreds of thousands of years. Some observers think that slight variations in the angle of the Earth's axis of rotation relative to the plane of its orbit round the sun, coinciding, as they very rarely do, with changes of distance from the sun, may change the received radiation from the sun enough to give the few degrees' change in average temperature of the Earth's surface that suffices to make a striking change in climate.

Apart from the changes in the Earth's distance from the sun, it is possible that the radiation from the sun may fluctuate. About every 11 years, on average, the sun develops a rash of 'sunspots', which are centres of disturbance at the edges of which extra brightness can be seen, indicating increased radiation. Sunspot activity has long been associated with climate changes on the Earth, such as cycles of drought in North America or of hot summers in Europe, but while changes in the heat supplied by the sun could be the reason for ice-ages, there is no firm evidence to point to this as the prime cause. If such a link were ever established, it would still leave us in the dilemma of having to forecast the sun's future activity in place of forecasting the Earth's climate.

At the present time, it is most important that increased effort be made to learn what we can from historical records of the weather and climate, so that pointers to future long-term changes can be recognized. As Dr B. J. Mason, Director-General of the United Kingdom Meteorological Office, explains:

Unfortunately our understanding of the mechanisms and cause of climatic trends and fluctuations is inadequate to allow of their prediction. It is not even clear whether they are brought about by internal changes in the atmospheric system or by changes in external facts such as the sun's radiation. The problem may in the foreseeable future be complicated further by the possibility that human activities, through the production of additional carbon dioxide, aerosols, and waste heat, may cause inadvertent but significant changes in weather and climate.*

---

* *Endeavour*, Vol. XXXV, No. 125, May 1976.

2 The weather machine

Climate means different things to different people, depending upon where they live. In warm tropical latitudes climate means a dry season and a wet season, each characterized by a dominant pattern of weather which may be easily predictable from day to day. On the other hand, in more temperate latitudes climate involves significant temperature changes from season to season and rainfall, or lack of it, may easily occur at any time with considerably less predictability than in tropical latitudes. Nevertheless climate really means something to most people and this is because, despite its capacity for variation, weather tends to have some common pattern during the same certain weeks or months or seasons of the year, recognizable from year to year and sufficiently distinguishable from other seasons of the year.

The two most important characteristics of climate are temperature and moisture. A third is atmospheric pressure, the distribution of which determines wind speed and direction. In order to understand the variations in dominant weather patterns across the world and through the year it is important first to separate cause and effect.

The most important basic causes are the amount of incident solar radiation and outgoing terrestrial radiation, which depend upon latitude and season. The Earth rotates around the sun once a year in a roughly circular pattern called the 'plane of the ecliptic'. At the same time the Earth is spinning about its own axis once every 24 hours. This axis itself is inclined at an angle of $66\frac{1}{2}°$ to the plane of the ecliptic, which means that as the Earth rotates around the sun there is an unequal and varying distribution of sunlight falling upon the Earth. When the Earth is at one side of the plane of the ecliptic (21 June) the North Pole is tilted towards the sun and the northern hemisphere receives its maximum amount of solar radiation while the southern hemisphere receives its least. Indeed, on this day all parts of the Arctic north of $66\frac{1}{2}°$ N have 24 hours of daylight while the Antarctic south of $66\frac{1}{2}°$ S has 24 hours' darkness. When the Earth is on the opposite side of the ecliptic plane (21 December) exactly the opposite situation occurs, with maximum daylight in the southern hemisphere and least sunlight in the northern hemisphere. On two other days in the year (21 March and 21 September) the Earth lies either side of the sun midway between the positions of maximum and minimum solar radiation, when the length of the day is the same everywhere around the world.

The temperature of the atmosphere is largely determined by the amount of incoming solar radiation offset by the outgoing radiation to space. The solar radiation heats the Earth's surface and this heat is transferred to the atmosphere by conduction and convection. The amount of heat received at any spot on the Earth's surface depends upon the angle of incidence of the solar radiation. When the sun is directly overhead (as it is at the equator on 21 March and 21 September, on the tropic of Cancer on 21 June and the tropic of Capricorn on 21 December) the solar radiation is absorbed directly by the Earth with maximum heating for a given area. On the other hand, in polar latitudes the angle of incidence is large and the same amount of radiation is spread over a greater area of the Earth's surface with a corresponding diminution of heat. This is the reason why the tropics are warmer than the poles despite the greater amount of daylight at the poles during their respective summers. Taking the year as a whole, therefore, there is a strong net gain of heat in equatorial latitudes and a strong net loss of heat in polar latitudes.

The next most important cause is the nature of the underlying surface. For example, land and sea have quite different specific heats; the same amount of heat will raise the land temperature more quickly than the sea temperature, but the land will give up its heat more easily than the sea for the same reason. Ice, snow and deserts are strong reflectors of sunlight and also radiate heat themselves quite strongly to space. Mountains, too, act as barriers to the movement of warm air and indirectly affect the distribution of rainfall.

Temperature is related to pressure by virtue of the fact that warm air is less dense than the cooler air surrounding it and hence exerts less pressure. Thus warm areas tend to be associated with low pressure and cold areas with high pressure.

Because pressure varies from place to place the air tends to move from high- to low-pressure areas in order to create equilibrium. This air movement is the familiar wind, but, because of the rotation of the Earth, the wind does not move in a straight line from high to low. At the equator the surface of the Earth is moving at about 1,000 miles an hour (the Earth's circumference is about 24,000 miles), while at the poles the speed is zero, although anyone standing at a pole will rotate once every 24 hours. If an object, be it a cricket ball, an artillery shell or a mass of air, moves away from the equator, it will be travelling west to east at 1,000 miles an hour in addition to its speed away from the equator. When it lands it will be at a place where the Earth's west–east velocity is less than 1,000 miles an hour, so the object will land to the east of where it was aiming when it left the equator. This 'kick' to the east can be represented mathematically as a force, analogous to centrifugal force, which is known as the Coriolis force. When an object approaches the equator the Coriolis force pushes it to the west, and

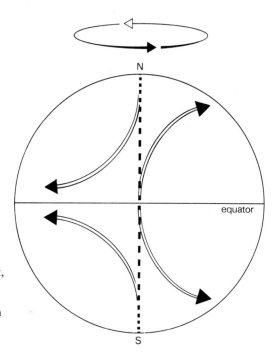

1 The Coriolis effect, as the earth spins, gives a kick to the right in the northern hemisphere, to the left in the southern

in general the rotation of the Earth turns straight-line travel into a curved path. In the northern hemisphere there is, then, a tendency for moving objects to deflect to the right of their line of movement. For example, the northward-moving Gulf Stream in the North Atlantic does just this, moving along the east coast of the USA, across the Atlantic and back south to the equator. In the southern hemisphere the southward-travelling counterpart travels anti-clockwise down the South American coast, across the South Atlantic and up the coast of Africa, trending left all the time. The wind spreading out from a high-pressure area in the northern hemisphere, because of the Coriolis force, travels in a clockwise direction rather than radially outwards, and in an anti-clockwise direction in the southern hemisphere. With low-pressure areas, the winds aiming towards the centre of the low still tend to the right in the northern hemisphere, and this means an anti-clockwise circulation around the low-pressure area rather than a direct impingement onto it from all directions.

Two more factors influence the air motion round the pressure system. Centrifugal forces cause the air to accelerate around high pressure and decelerate around low pressure when the path of the air is strongly curved. Friction between the air and the Earth's surface causes the air to be deflected inwards towards low pressure and outwards away from high pressure. This is because the friction reduces the air speed, which reduces the Coriolis effect and allows the air motion a greater component towards the lowest pressure.

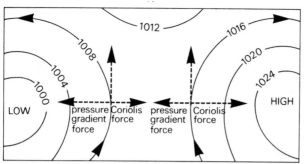

2 The Conolis effect and pressure gradient force; air moves from high to low pressure and as a result the winds blow clockwise round anticyclones (high) and anti-clockwise round depressions (low)

This means that the effect of friction is to cause a gain of air within low-pressure centres and a loss of air within high-pressure centres – that is, a tendency to weaken both systems.

Low-pressure centres are commonly called depressions, or troughs; in tropical areas very intense depressions are called cyclones, hurricanes and typhoons. High-pressure centres are called anticyclones, or ridges. Anyone familiar with looking at weather maps soon notices an important difference between depressions and anticyclones in that the pressure changes are usually much stronger around depressions than anticyclones, as shown by the isobars being closer together. This is because of the need for balance between the pressure difference, the Coriolis force and the centrifugal forces in the air motion. For anticyclonic flow there is a definite limit to the rate of change of pressure with distance near the centre of the anticyclone, whereas for cyclonic flow there is no limit to the steepness of the slope of pressure. Thus winds are usually light in anticyclones but may be very strong in depressions.

Since winds blow in circular patterns instead of straight from high- to low-pressure areas and since, in general, pressures are high in cold and low in warm parts of the Earth, one might expect that where, for example, winds blow from the north-east, there would be warm air to the south-east and cold air to the north-east in the northern hemisphere. Unfortunately these relationships are only found in the lowest layers of the atmosphere and in certain areas. Both pressure and temperature decrease much more rapidly in the vertical than in the horizontal plane. In general the rate at which the temperature falls (called the 'lapse' rate, presumably because it is a decline to a lower state) is about 5°C per kilometre up to about 10 km, above which the temperature levels off or even begins to rise again. The level at which the lapse rate changes abruptly is called the tropopause and separates the troposphere from the stratosphere. The tropopause, which is located by meteorological balloons carrying temperature-measuring equipment, may vary in height from 5 km in arctic air to over 12 km in tropical air.

Pressure drops to half its surface value at about 5 km and to less than a third at about 9 km. The rate of fall of pressure with height is determined partly by the rate of fall of temperature. For instance, if over an area of the Earth's surface part of the area is covered by a cold air mass and another part is covered by a warm air mass, owing to the differing densities in the adjacent air masses the pressure will decrease more rapidly with height in the cold air than in the warm air. Thus at any fixed height above the Earth's surface there will be horizontal changes of pressure and air will move under the same dynamical constraints as at the surface around depressions and anticyclones.

There is, however, an important difference: because upper lows are generally cold and upper highs are warm (the opposite to the association of low pressure with warm at the Earth's surface), upper westerly winds are usually associated with cold air to the north and warm air to the south. As the difference of pressure between the adjacent warm and cold air masses becomes accentuated with increasing height the strength of the upper winds is found to increase with height, in fact reaching a maximum just below the tropopause where the rate of fall of temperature changes significantly. The strongest winds at these upper levels are called jetstreams; their speeds are commonly over one hundred miles an hour and occasionally are double this value. The term 'jetstream' seems to have been coined by Professor Rossby in 1946 for the general belt of upper-atmosphere winds which blow in both hemispheres in temperate latitudes, but today 'jetstream' is applied only to the strongest winds, which extend much further in the direction of the wind than they do laterally. Although the winds are important for modern air travel the name has no significance in this connotation.

The actual pressure at a fixed height depends in part upon the pressure at the Earth's surface as well as on the density structure in the layer between. Thus the high pressure at the surface within the comparatively steady anticyclone that sits over Siberia in winter is replaced by relatively low pressure at upper levels because of the presence of very cold air in the intervening layers. It is therefore standard meteorological practice to analyse upper wind systems along fixed pressure surfaces rather than at fixed heights, because in this way the relationship between adjacent pressure surfaces in the vertical depends entirely on the temperature structure and not on density. For example, if the height of the 1,000-millibar surface is zero (i.e. the surface pressure is 1,000 mb) the height of the 500 mb surface will vary only as the mean temperature of the 1,000–500 mb layer since the temperature in this interval determines the air density which provides the drop of 500 mb. The same applies to any pressure level above 1,000 mb and the higher we go in the troposphere the closer is the identity of the contour surface with the mean temperature in the air below. For this reason the jetstreams found in the upper troposphere which are caused by the horizontal pressure changes in the upper atmosphere, are closely identified

The weather machine

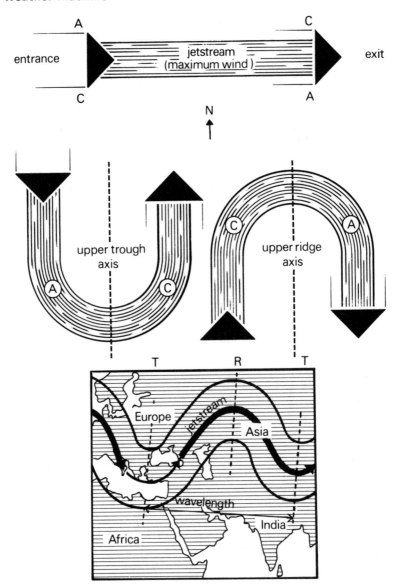

3 The regions of development of depressions and anticyclones near the surface of the earth in relation to the jetstreams at the upper levels in the troposphere. Note the almost horizontal movement of air at jetstream levels and the scale of the horizontal movement of air compared with the vertical depth. (C: depression development, A: anticyclone development, T: upper trough axis, R: upper ridge axis)

with the zones of strong temperature contrast in the middle and lower troposphere. These zones are called 'frontal zones' and separate air masses of different heat and moisture content.

The temperature near the Earth's surface does not change uniformly between the equator and the poles. Instead the distribution of land, sea and mountains causes air masses to acquire varying homogeneous proportions of temperature and moisture content over extensive areas where the air stagnates for periods of a week or more. For example, air that stagnates over the Sahara becomes hot and dry while air over the tropical ocean becomes very warm and damp. There are numerous types of air masses which are classified on the basis of their origin, ranging from tropical to Arctic and continental to maritime.

Air masses are modified by the underlying surface when they leave their source regions. Arctic continental air often sweeps across the north-west Atlantic from its source over Canada in winter. The warm waters supply copious heat and moisture to the air mass by vigorous convection so that by the time the air reaches western Europe it has become polar maritime air, characterized by frequent showers and thunderstorms. Conversely, when tropical maritime air sweeps northwards across the Azores to north-west Europe the air is cooled in its lowest layers by contact with a progressively cooler ocean. The cooling causes the original very damp air to become saturated and allows the formation of thick low cloud and sea fog. This air mass may lose its excess moisture on crossing coastal areas of western Europe so that well inland the low cloud often disperses and warm sunny weather develops.

Although the Earth's characteristics allow air masses to develop with quite sharp boundaries close to the physical boundaries such as coasts, the atmosphere itself has the capacity to intensify or destroy the temperature differences as the wind fields move through the boundaries. These thermal boundaries or fronts appear to be an intrinsic characteristic of the atmosphere's motion. Fronts usually move and evolve in a fairly systematic manner. Initially the warm air mass begins to move north by climbing above the cold air to the north. The sloping front, which itself moves slowly north, is called a *warm front*. Later the cold air mass sweeps southwards to the west of the warm front and undercuts the warm air mass. This sloping front is called a *cold front* and the slope is usually much steeper than the warm front. Subsequently the cold front overtakes the warm front so that all the warm air is lifted above the surface. The new front is called an *occlusion*. In fact the occlusion may not represent a temperature boundary at all if the cold air on either side is, by and large, of the same origin. Often, however, the air masses are complex and the air behind an occlusion may be colder or warmer than the air ahead. In such cases the front is called a cold or warm occlusion respectively.

Fronts are usually associated with active weather, depressions, rain, storms and gales, but not necessarily so. A moving front may be marked by nothing more than a thin sheet of cloud or even by the appearance of some cumulus clouds. To understand these variations of activity in fronts we need to know more about the causes of weather.

Precipitation (i.e. rain, hail or snow) occurs predominantly as a result of two distinct processes in the troposphere. In one process, called *convection*, individual 'parcels' of air are heated at the Earth's surface (either by the sun or a warm ocean) and rise through a colder ambient air mass. The air parcel cools owing to expansion (lower pressure aloft) and eventually stops ascending when its temperature is the same as the ambient temperature. Before this happens the air parcel may have cooled to the stage at which it cannot hold its supply of moisture without condensation in the form of cloud. Once the cloud has formed there is a release of latent heat which gives more impetus to the ascending air parcel. Further ascent eventually leads to massive cloud growth with large, heavy drops of water which fall to the ground as rain. This sort of rainfall occurs in showers and thunderstorms and the rainfall can be very heavy as a result of the strong ascent of the air. However, because it is individual air parcels that have risen, compensating air parcels are forced to descend to replace them. Descending air warms by compression and evaporates moisture droplets so that regions of descent are usually characterized by clear skies. Thus convective clouds tend to be cellular with clear areas between them. Convective rainfall occurs almost anywhere in the world but most of all in tropical and subtropical regions where surface heating of the transient air masses tends to be greatest.

The second most important cause of precipitation is the process of *upslope* motion. Here the entire section of the air mass is ascending (less rapidly than in the convective process) without local compensating descent; the cloud structure tends to be more extensive in area, and also layered, and the rainfall is generally less intense than the convective sort. What causes upslope motion? The answer lies in a fundamental atmospheric relationship called the 'equation of continuity', which simply states that if air is being added or removed at any level within a region of the atmosphere there must be compensating movement of air to or from different levels within the same region. The addition and removal of air (called convergence and divergence respectively) occurs at any level but most often in the lower and upper parts of the troposphere with a distinct minimum, usually in mid-troposphere, i.e. about 500 mb. Thus if there is convergence near the Earth's surface and divergence at jet-stream levels the air mass ascends to satisfy continuity. Reverse the distribution of convergence and divergence and the air mass descends for continuity. Divergence (and convergence) is usually

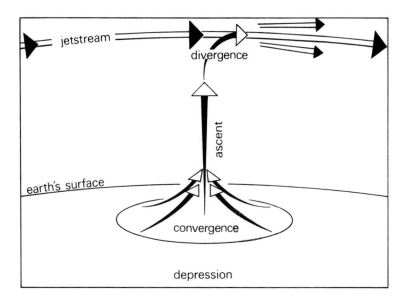

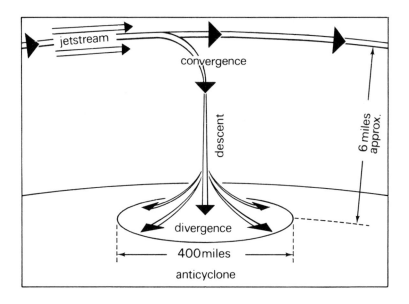

4 Development of a depression (*top*) and an anticyclone. Divergence at upper levels causes a loss of air. This is replaced by ascending air, which is fed by convergence at the surface. Convergence of air aloft causes accumulation of air, part of which descends and diverges at the surface

greatest in the upper layers rather than in the lower layers. It follows that upslope motion in the air mass tends to be controlled by events near the jetstream level in the upper troposphere. We shall return to this theme shortly in studying particular climatic regimes.

Divergence and convergence in the upper troposphere are also closely related to the development of depressions and anticyclones, which usually form near the Earth's surface and gradually extend upwards with time. If there is strong divergence in the upper troposphere within a certain region, the loss of air in the column is reflected by a drop of pressure at the surface and the development of a depression. The loss of air is partially offset by convergence near the surface and ascent within the column and in point of fact the full development of the depression represents the net loss of air associated with large opposing divergence and convergence fields within the air column. It is also clear that depressions are associated with extensive cloud and precipitation because of the ascending air. Anticyclones develop when strong convergence is present near the jetstream level and there is an accumulation of air within the column with descent for continuity. Accordingly anticyclones are usually associated with dry weather and often with clear skies. The development of a depression and an anticyclone is shown schematically in Fig. 4.

It will not have escaped the discerning reader that the close relationship between the jetstream and strong horizontal temperature changes below means that upslope and downslope motion and the development of depressions and anticyclones occur precisely within these zones of strong temperature contrast or fronts.

The restless motion of the atmosphere, with its bringing together of air masses of sharply differing origins, can produce some fearful storms, ranging from the single-cloud thunderstorms, possibly accompanied by a tornado, to a tropical hurricane with a diameter of some 300 kilometres, and also the intense depressions such as are found near Iceland and the Aleutians, which are as much as 1,000 kilometres across.

Thunderstorms are an extreme result of vigorous, deep convection in the troposphere, when the clouds build up to a height often in excess of 9 or 10 km. There is a great mass of water and ice carried within the clouds and supported by violent upcurrents of air. Subsequently the air currents release the water and ice and phenomenal falls of rain result, often accompanied by hail. The alternate freezing and melting and breaking up of the water drops within the clouds leads to an enormous build-up and separation of electrical charge inside the cloud. The voltage build-up is released by lightning discharge from cloud to cloud as well as from cloud to land. Although thunderstorms can occur anywhere they are most common in the tropical areas where heat and moisture are plentiful at the Earth's surface.

When the ascent of warm, moist air is very rapid an upward spiral movement sometimes develops, and quickly builds up to a narrow funnel of air whirling at a speed of more than 600 kilometres per hour. Condensation of water is rapid within the funnel which extends down from the cloud base to the ground. This terrifying phenomenon, the tornado, sweeps across the land like a vacuum cleaner at speeds varying from 40 to 60 km per hour. It is some small comfort to know that tornadoes rarely last longer than about eight minutes and are only 20 to 300 metres across. Tornadoes are most common in the Mississippi–Missouri basin, where there is often a conjunction of hot moist air from the Gulf of Mexico with cold Arctic air sweeping south from the prairies. Tornadoes can cause tremendous loss of life and property; on 11–12 April 1965, 47 tornadoes in the Ohio Valley caused 257 deaths and damage estimated at 200 million dollars.

The most destructive kind of storm is the hurricane or typhoon which develops over tropical waters not too close to the equator. These tropical storms are intense wind circulations with a centre of low pressure; they owe their origin to a combination of upslope motion in the equatorial winds and convection from the ocean with a sea surface temperature above 27° C. Hurricanes have an average life span of about six days from birth to reaching land or temperate latitudes. Their evolution can be divided into four stages. The *formative* stage may take several days, but can also be very quick with the strongest wind usually concentrated in one quadrant and below hurricane force (which is about 75 m.p.h.). During the *immature* stage winds of hurricane force form a tight band within about 30–45 km of the centre and the pattern of cloud and rain becomes organized in spiralling bands, but the overall area of the storm is still small. In the *mature* stage the maximum winds are no longer increasing but the storm increases in area up to 300 square km or so although there are marked variations. Later the hurricane reaches the *decaying* stage, usually when the storm curves round from the tropics and enters the belt of westerly winds in temperate latitudes.

No one knows how high wind speeds can rise in tropical storms but speeds above 160 m.p.h. have been recorded. Sustained wind speeds of 120 m.p.h. occur over a limited area and as the pressure exerted by wind on buildings is proportional to the square of the wind speed it is not surprising that few buildings survive the passage of a hurricane. In the course of a hurricane's life a vast quantity of air is drawn into the circulation and funnelled upwards with a colossal release of latent heat, estimated at from 2 to 4 times 10^{26} ergs per day, which, in simpler terms, is the equivalent of burning 20,000 million tons of oil. Since it is observed that only some 15–20 per cent of this heat energy is required to maintain the circulation of the storm, then clearly the rest of the heat must be extracted and used to heat vast areas of the upper troposphere away from the hurricane's location. In this sense, therefore, hurricanes perform a useful role in transporting heat from the tropics to temperate latitudes.

There is one more particularly important aspect of weather to describe: this concerns the relationship between patterns of weather around the globe. Despite the presence of upward and downward motion the movement of the air in the troposphere is, broadly speaking, horizontal. The air currents in the middle and upper troposphere meander around the hemispheres in waves, rather like ripples on a rope, sometimes having large amplitudes and sometimes having small deviations from the latitude circles. Now it is an intrinsic law of air motion that the horizontal distance between the troughs and ridges in the wave train is directly related to the strength (or speed) of the air flow and also the amount of curvature in the flow (or the sharpness of the troughs and ridges). Also the speed of motion of a wave train in the flow itself depends largely upon the wavelength, curvature and air speed within the wave. Basically what we find is that when highs and lows are close together they tend to move rather quickly, perhaps 30 to 40 knots, whereas when they are far apart their speed of movement is very slow, about 5 to 10 knots. So we see that the atmosphere is in a continuous churning motion with small eddies being flung off as pockets of high or low pressure, but with underlying steady movement in certain directions. Like an infantry battle in the days of hand-to-hand fighting, small groups in their individual mêlées appear most important to those thus engaged, but the overall picture is of two great opposing movements. No one can tell the result of each little sub-unit of the battle, but in the long run it must be these that determine the outcome of the battle. Sometimes the success of one group spurs neighbouring groups to follow suit. The unexpected may occur with the arrival of Blücher and large reinforcements.

The Earth's atmosphere is held to the Earth by gravitational attraction; it is warmed by the sun from outside and it is both heated and cooled by the land and the sea from below; all the time the Earth rotates and drags the air with it, and water evaporates, precipitates, freezes and melts to complicate the picture. It is not possible to leave a vacuum in the atmosphere, so if one mass of air moves up a body of air must come down somewhere to replace it. The jetstreams of the upper atmosphere have their own little regime, but they are neighbours to the air below which we breathe, and so what happens at 20,000 feet up is also a contributor to our own local weather. It is not surprising that the weather is complicated, changeable and difficult to forecast.

3 The six seasons

Just as there are several kinds of weather phenomena, ranging from thunderstorms to Icelandic depressions, so there are scales of climatic regimes within which weather patterns evolve in a partly ordered way. The regular apparent movement of the overhead sun from the equator to the tropic of Cancer and back again to the tropic of Capricorn causes a similar movement of the hot tropical air masses. Similarly the continents become very warm in summer and very cold in winter. These regular features of climate impress themselves strongly upon the local weather, and indeed if there were no other influence the climate would simply be measured by the amount of heating and associated convection or the amount of radiation cooling. The essential point is that the regular features of climate are always there but their influence upon the weather depends upon how active are such elements as depressions, anticyclones and fronts in the daily evolution of the atmosphere.

There are several incontrovertible reasons why the global atmospheric circulation should have a basic ordered system which transcends the daily variability of weather. The first reason is the heat balance, since equatorial regions receive an excess of incoming solar radiation over outgoing terrestrial radiation while polar regions have the reverse distribution. The equatorial atmosphere does not continuously rise in temperature nor does the polar atmosphere continuously cool, so there must be some mechanism for transporting the heat polewards in a regular manner to keep the temperature roughly constant. Secondly, the Earth's rotation rate is not observed to vary significantly. This implies that westerly winds, which would tend to accelerate the Earth's rotation, must be balanced by easterly winds which would tend to slow it down. Thirdly, the long-term distribution of water vapour is roughly constant despite the varying capacity for evaporation over the oceans and condensation through rainfall. Fourthly, over the centuries there is no significant variation in the air pressure, which implies that, in the long term, there is a steady distribution of the mass of air around the Earth.

In order to balance its budget the atmospheric circulation has certain well-ordered types of evolution. The hot equatorial air rises by convection and moves northwards in the upper layers (in the northern hemisphere). At about $30°$ N part of this warm air descends to the lower layers and returns southwards as the north-east trades towards the equatorial low-pressure area, thus completing

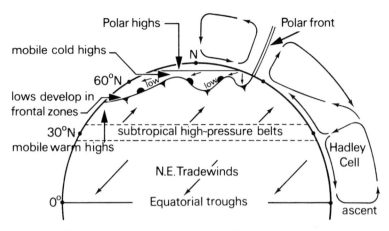

5 A typical pattern of winds over the northern hemisphere

a north–south circulation regime called the Hadley cell, after the eighteenth-century pioneer of meteorology who first described it. Part of the warm air, however, continues to move northwards to middle latitudes where it helps to maintain the temperature changes in these areas. In polar regions high pressure predominates at the surface, associated with the very cold and therefore heavier air. The cold air moves away from the polar cap and adds to the temperature changes in middle latitudes. The mechanism for completing the transfer of warm air northwards and cold air southwards comes through the development of depressions and anticyclones in the strong frontal zones found in middle latitudes.

So, despite the variability of weather from day to day, it is all part of a grand plan. The thunderclouds of the tropics carry the heat and moisture upwards and northwards, occasionally aided on a larger scale by hurricanes. The semi-permanent Icelandic and Aleutian depressions, familiar to those who follow the weather forecasts, maintain the northward transport of heat and moisture to the polar regions while driving the cold air south to the subtropics and reinforcing the north-east trades to complete the cycle.

So much for the overall features that exert a controlling influence upon climate and weather. Let us examine in more detail some climatic regimes that are found over the northern hemisphere.

A comparison of the mean charts for January and July easily reveals significant differences in temperature and rainfall. Figs. 6 and 7 illustrate the distribution of temperature in January and July respectively. It is apparent that in winter the lowest temperatures are found over the land and the seas are relatively warm, whereas in summer the land tends to be warm relative to the sea. There are other interesting features of the temperature distribution: whereas the summer minimum is centred over the Arctic ice the winter minimum is centred over two areas, northern

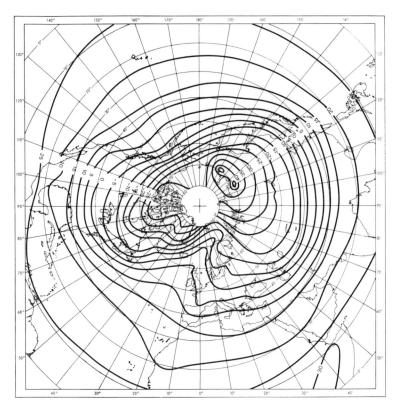

6 Mean temperature (°C) over the northern hemisphere in January

Canada and north-east Asia. Notice, too, how the Rockies cause the winter isotherms to dip southwards east of the high ground and also cause the summer isotherms to peak northwards with a pronounced hot cell over the Grand Basin. However, one of the most important features is the distribution of the zones of strongest temperature change. For example, in January a rapid hot-to-cold trend extends from California across the southern USA to the south-east of Newfoundland. To the east there are two different zones, one extending north-east to the east of Greenland, to Spitzbergen and thence south-eastwards to north-west Russia, and another spreading eastwards south of the Azores to North Africa, Arabia and southern Russia. It can be seen that the southernmost zones of strong temperature change tend to shift northwards between winter and summer before retreating southwards again.

The greatest rainfalls are found in tropical latitudes, simply because warm air can hold more moisture than cold air. Nevertheless, dry regions *are* found in low latitudes, mainly over North Africa, Arabia, and southern Asia north of the Himalayas, and also east and south of the Rockies. Indeed the Rockies are a mighty influence on the local rainfall: copious amounts fall on the coastal

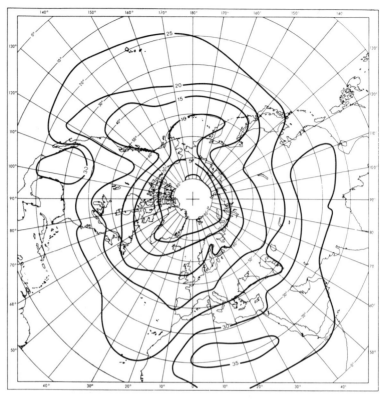

7 Mean temperature (°C) over the northern hemisphere in July

side of the range, where moist oceanic air is forced to ascend, but much less rain falls on the leeward side of the range. It is noticeable that, at least in lower latitudes, the rainfall patterns tend to move northwards between winter and summer.

The annual trend of the atmospheric circulation has been studied by many climatologists and the delineation of seasons based upon 'natural regimes' in the circulation has tended to favour six seasons rather than the conventional four. The concept of six natural seasons has found most favour with Russian meteorologists. A similar division has been found useful for some purposes by the United Kingdom Meteorological Office.

The six seasons are illustrated in composite form by the distribution of surface pressure, 500 mb jetstreams and the surface–500 mb mean temperature or 'thickness'. The thickness of the lower layer of the atmosphere is a sensible measure because it gives one an idea of the average temperature of the air: if it is hotter the air is lighter and so the layer is deeper.

In the description of each season certain mechanisms are postulated to explain the interseasonal changes in specific localities in terms of the broadscale circulation.

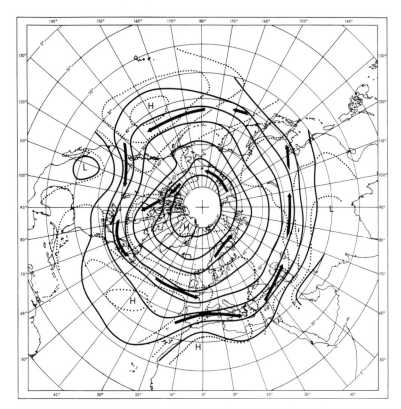

8 Distribution of surface pressure, 500 mb jetstream, and thickness (mean temperature) – in other words, the average weather pattern – of the northern hemisphere in summer. Continuous lines represent the lines of equal thickness, dotted lines the isobars at mean sea-level and thick arrows the jetstreams

Summer (July–August)

Summer in fact marks the transition from heating to cooling at all latitudes, with the transition most rapid in the north. Notice (Fig. 8) how the coldest air is located over the Arctic with marked extensions over the north Pacific Ocean, Labrador, Norwegian Sea and north central Siberia. The most prominent hot zone is over the western USA but other hot areas are south of the Atlas Mountains, Arabia and much of southern Asia. In summer the hot centres are at their most intense and consequently have their greatest influence upon the atmospheric circulation. In particular the location of the hot centre just east of the southern Rockies influences the position of the downstream cold trough over the St Lawrence estuary, which in turn influences the downstream warm and cold formations, the latter across the United Kingdom. Indeed, the onset of the traditional British summer weather, consisting of rather cool

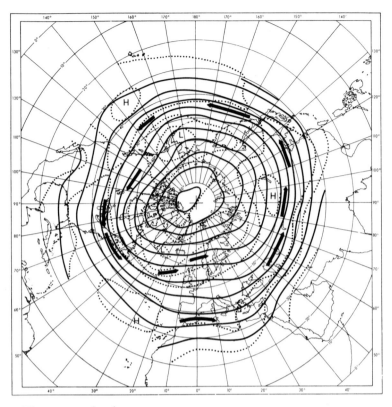

9 The pattern develops: autumn

oceanic rain-bearing westerly winds, can be linked to the establishment of the intense hot high-pressure ridge over the western USA. This indirectly explains why the amateur forecaster believes that the UK gets the same sort of weather that the eastern USA had last week. The surface pressure is dominated by large anticyclones over the east Pacific and the Azores with extensive westerly winds on their northern flanks, while low pressure is closely identified with the hot areas over the land in low latitudes. Summer is the season of overall small temperature changes, and therefore is the season of weakest cyclonic and anticyclonic developments; thus with the hot centres controlling the positions of the areas of cool and warm around the hemisphere, summer tends to be a season with a persistent type of weather.

Why therefore do we have different types of summers from year to year? The reason is that quite small variations in the mean position of the atmospheric systems are sufficient to create very different local weather. For example, most of the best summers in north-west Europe occur when there is a fairly steep change of temperature over the Atlantic, shown near 50° N in Fig. 8, which shifts some 5° or so further north across the north-west seaboard of Scotland. Similar shifts southwards towards the Azores usually result in poor summers over much of north-west Europe.

These shifts in the zones of strong temperature change may be a result of particular developments in the circulation elsewhere, such as over Canada or Scandinavia, but possible causes may be linked to relatively high or low sea-surface temperatures whereby the exchange of heat from sea to air generates temperature changes in the lower atmosphere. Some sceptics say that if a polar bear sneezes in Greenland there is a strong possibility of snow in Africa!

Autumn (September–October)

Autumn is the season of most rapid cooling generally but there is a tendency for a distinct southward movement of the cooling maximum from the polar region in September to middle latitudes by the end of October.

A comparison of the autumn mean thickness of the atmosphere up to the jetstream layer with the summer thickness reveals a considerable increase of temperature change around the hemispheres, mainly as a result of rapid cooling over eastern Siberia and Canada. Notice how the summer hot zones have lost their intensity while the intensifying cold areas cause a marked rearrangement in the position of the pattern of temperature and in particular a marked shift in the location of strongest temperature changes from the east Atlantic to the Norwegian Sea.

The surface pressure field shows considerable development of the Aleutian and Icelandic depressions compared with summer and a corresponding weakening of the subtropical anticyclones. Autumn therefore differs from summer chiefly by the presence of intensifying storms in high latitudes.

At the same time there is usually a marked improvement in the character of the weather over north-west Europe, brought about by the appearance of large anticyclones which develop on the warm side of the steepening Arctic temperature 'slope'.

Prewinter (November–December)

Rapid cooling continues overland during prewinter together with intensification of the temperature differences, as might be expected, around the flanks of the two major cold centres over Manchuria and eastern Canada. The strong changes are located further south than during autumn, especially over the continents and western oceans. However, what really distinguishes prewinter from autumn, in addition to the further advance of cooling, is the tidying-up of the high- and low-pressure areas, which are put into their place by the major cold air masses over eastern North America and

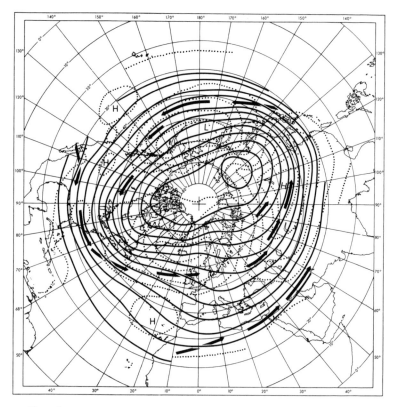

10 Prewinter

east Asia. A ridge of high pressure develops across the central Atlantic, linking the U S A cold mass to a similar European feature, while over the Pacific sector the intense broad Siberian cold mass absorbs the hitherto autumn Aleutian cold mass and forces up the Rockies warm mass into the Gulf of Alaska. These 'cause and effect' thermal adjustments manifest themselves in the mean surface pressure fields. Both the Aleutian and Icelandic depressions have intensified and broadened and a copious supply of oceanic heat has had its effect, while there is an increased likelihood of depressions over southern Europe, particularly over the Mediterranean Sea where the warm waters help to intensify the storms.

Winter (January–February)

Winter is the season when overall cooling comes to an end and the major low-temperature areas have extended to their limits southwards across the continents. Because corresponding cooling has not occurred over the oceans, the temperature lines (or iso-

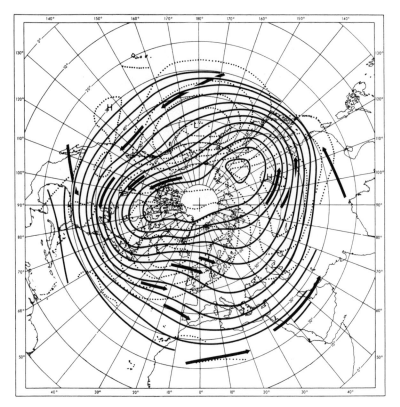

11 Winter

therms) attain their maximum north–south inclination in winter.
This means that although the winter Aleutian and Icelandic depressions are still very intense, there is a marked tendency for large anticyclones to develop downstream of the depressions, as over the
Rockies and western Europe, with their centres usually in high latitudes. When these anticyclones develop they bring an interruption
to the normal mild westerly winds flowing into western Europe
and the Rockies and, over Europe in particular, the anticyclones
may link with the intense monsoonal Siberian winter anticyclone
and cause very cold air to sweep westwards to the European seaboard.

As the intense cold air is swept south and east across the western
Pacific and western Atlantic it has the effect of increasing the temperature contrasts which extend across the more southerly parts
of the oceans and attempt to link with the temperature changes
over the extreme southern Rockies and North Africa respectively.
These developments allow secondary depressions to move eastwards across the oceans well south of their earlier routes.

33

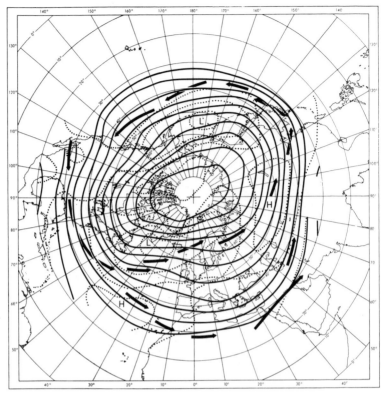

12 Spring

Spring (March–April)

Heating develops strongly during spring although it does not become well marked in the Arctic until April, because the ice reflects so much of the sun's heat. Heating is in fact most rapid in middle latitudes within the continents, while insignificant temperature changes are observed over the oceans. Although this process gradually reduces the north–south amplitude of the isotherms, the actual distribution of heating leads to a rearrangement in the development of weather systems. The Aleutian and Icelandic depressions are still present but tend to be weaker because the relatively cold ocean provides less sensible heat to intensify the storms. At the same time the strong heating in the subtropics causes depressions which develop over the Gulf of Mexico and migrate eastwards, occasionally turning north-east to engage the Arctic air from Canada and become major storms near Greenland. Some of these depressions move across the Atlantic close to the Azores towards north-west Africa, where the strong heating is already helping to develop depressions over Algeria. These desert storms move east towards Egypt and Arabia and are responsible for the extensive dust storms called *khamsin* or sirocco.

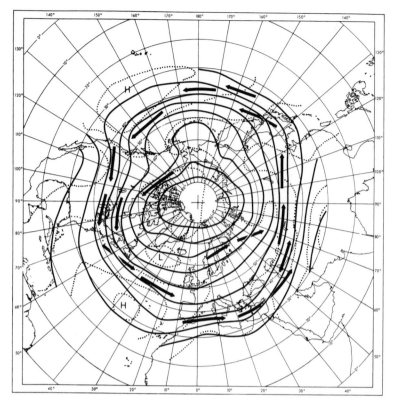

13 Presummer

Presummer (May–June)

Presummer is the season of strongest heating in all latitudes, with maximum values in the polar region. The heating is still most evident over the continents. The distribution of heating leads to a smoothing of the temperature pattern in most sections of the hemisphere. The hot regions are beginning to exert their influence over the western USA, north-west Africa, Arabia and southern Asia generally. Although the coldest air is found within the Arctic there are relatively cold areas over both oceans which illustrates how the ocean heating lags behind the continental heating. The temperature differences are less, and therefore cyclonic and anticyclonic activity is also less intense, in both high- and low-latitude streams. In presummer the residual cells of cold Arctic air tend to stagnate over the oceans and allow anticyclones to develop and persist at times over the eastern seaboards, as is found in western Europe and the northern Rockies. So this season really signifies the end of the process of warming the long winter accumulation of cold air in the Arctic, while in subtropical regions the hot cells are developing and preparing to exert their controlling influence during the forthcoming summer.

It makes life more interesting to chop the year up into six pieces rather than the traditional spring, summer, autumn and winter, and it makes more sense meteorologically. However, there is no point in doing this if one forgets the underlying principles of the movement of the atmosphere and the troublesome effects of the jetstream and other influences on the body of air which exerts its pressure on us at sea-level.

Those unfortunate people – by and large, the leaders in meteorology, oddly enough – who live in the temperate zones suffer, on the whole, from the most miserable weather to affect this Earth. It may be a comfort to them to look back into history and join that strange band of scientists, the palaeoclimatologists.

4 Patterns of the past

John King, a preacher in Elizabethan times, was declaring in 1595, 'our years are turned upside down; our summers are no summers, our harvests are no harvests.' The river Thames froze over in London that winter and if John King had only known, things were to get worse rather than better, since a sequence of wet summers from 1594 to 1598 ruined the harvests and brought famine to Europe. The bad weather in what was popularly supposed to have been some of the best times to live in England was part of what those who study climate call the Little Ice-Age; when for a few hundred years the weather in Europe and America was considerably worse for man than it is today, although the land was not covered by ice-sheets as in the real geological ice-ages of the past.

The Little Ice-Age was at its worst in most parts of the world from 1550 to 1700, although some of the evidence points to an onset as early as 1310 in some places, and a reading of the world-wide coverage of observations of temperature and rainfall that is possible for the past 200–300 years suggests a later ending between 1850 and 1900. The trouble is that, although the conditions in the cold period were on average worse than at other times, there are always fluctuations, and good summers with plentiful harvests occur even in the most miserable periods of bad weather.

As might be expected, ice plays a large part in the change of climate at this period. Whether this is cause or effect remains to be seen when active research into climate changes reveals more of the truth of the mechanisms behind the large- and small-scale variations which have taken place in the past and which are, except for masterly intervention by man, bound to occur in the future. There is no doubt that during the cold centuries of the Little Ice-Age, the Arctic ice-sheet spread southwards. The old sailing routes to Greenland which had flourished in the Viking days gradually disappeared both on the eastern and the south-western shores of the island. Although ice had been reported by Irish monks in the sixth to the eighth century in the Iceland area, the Viking reports in the period AD 900–1200 mention difficulty from ice only a few times, yet by 1250 the ice is reported as formidable and by 1350 the direct routes from Iceland to Greenland had to be changed to avoid the ice. The ice at one time extended beyond Iceland to the Faeroes, which lie 200 miles north-west of the Shetlands, and even allowed a polar bear to land there from the ice-sheet.

A similar advance of glaciers in Europe is shown by reports, pictures and maps of the period, which tell of the cutting off of irrigation channels, and in one case the re-siting of a church to avoid destruction by the encroaching wall of ice. Precise data, by carbon dating methods (p. 59), of when trees were uprooted by the glacier show that the Grindelwald glacier was moving forward in AD 1280. After a pause of three hundred years, this well-recorded feature pushed forward and reached its most advanced position by about 1600. Sometimes, the glaciers would block valleys and form lakes, and historical reports give exact dates when catastrophe occurred with the ice dam breaking after heavy rains. Similar spreads of ice-sheets took place in Iceland during the Little Ice-Age. In Norway, glacier advances were destroying meadows before the beginning of the eighteenth century, with forward movements of hundreds of metres.

The Little Ice-Age was, then, a real southward spread of Arctic ice and may one day provide some clues as to the reason for the starting and stopping of ice-ages. In North America, the lack of a literate population in the early part of the period has precluded detailed evidence such as has been gathered for Europe, but the fact that the tree line moved downslope in the Rocky Mountains, as it did in central Europe, provides information of a southward movement of ice conditions. The one part of the world where there appears to have been less ice than the average was in the south polar regions. In the latter part of the Little Ice-Age, Captain Cook and other voyagers went further south than the present normal zone of pack ice, and with much less specialized ships than Antarctic explorers use today. In South America there is further evidence, from old rainfall observations, that the climatic zones of the Earth drift south and north together, so that if the Arctic ice moves down on Iceland, it recedes in the southern hemisphere.

All these clues to the behaviour of the world climate in the past are being collected by palaeoclimatologists, researching into old documents, maps and pictures which may provide facts, if only by inference, of what weather pattern existed in the past. The Little Ice-Age is important for these investigations for this reason. The whole episode took place when, in Europe at least, people kept records of population, taxes paid, imports and exports of wheat etc., and also made note of extraordinary happenings such as floods, crop failures, frozen rivers and other local catastrophes. All these records show to some extent whether it was hot or cold, and wet or dry, and since temperature and rainfall are the two most critical factors which determine climate, or what ordinary mortals call good or bad weather, it is possible to deduce from the written records a great deal about past climates. During the latter part of the Little Ice-Age, these deductions from old archives are paralleled by direct evidence from a gradually increasing, and now world-

4 A summary of climatic history. A: Changes over 100 years in the 5-year running average of surface temperature in the northern hemisphere, measured in tenths of a degree centigrade. (1) indicates the hot summers and mild winters of the 1940s. B: The winter severity index for eastern Europe over the last 1,000 years shows up the so-called 'Little Ice-Age' (2), about 1400–1700. The range from minimum to maximum is no more than 1.5°C. C: Shown in the broader perspective of 20,000 years the Little Ice-Age appears as a mere kink in the curve, which can now be seen to drop to 10°C below the present norm. D: The Younger Dryas cold interval (3) is a 5°C drop in a generally warmer trend, which is the present interglacial or Holocene (4), and we have to go back 125,000 years to find the last comparable warm age, the Eemian interglacial (5). E: From an examination of global ice volume it can be seen that over the last million years there have only been three warm peaks comparable to the present one. The scale shown represents a difference of about 50 million cu. km of ice, i.e. about 12°C

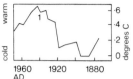

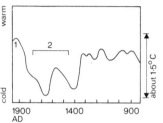

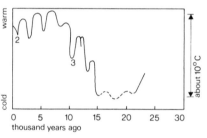

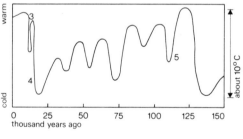

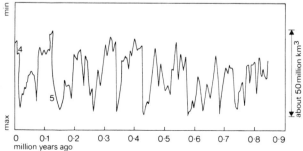

wide, network of meteorological observing stations, which have produced up to 300 years' continuous records of barometric pressure, wind speed, temperature and rainfall. It is from these observations that meteorologists draw world weather maps, and build up theories of how the atmosphere behaves, so that forecasts of future weather can be made, and it is from these regularly produced maps that our understanding of the mechanisms of the behaviour of the atmosphere will be obtained. The overlap of the modern measurements with the old documentary evidence makes it possible to construct weather maps for times in the past, and so to connect changes in climate with changes in the average circulation of the air.

In a later chapter, we shall see that there are other ways, such as analyses of tree-rings, examination of deep ocean sediments and cores from polar ice-caps, observation of sea-level changes, as well as modern isotope methods of determining the age of various materials, which can help us put numbers to some of the deductions made from historical research. As Professor Lamb points out, in *Climate: Present, Past and Future*, these new methods are most valuable since there is a limit to the climatic information that can be derived from history. For example, the analysis of wheat prices, which should indicate shortages and hence bad seasons, is dependent also on the amount of monetary inflation. The discontinuing of grape-growing in England probably provides a definite indicator of a worsening of climate, but fluctuations in the wine harvest, of which there are centuries-old records in some parts of Europe, may depend on wars or pestilence rather than on climate. On the other hand, there are some cases where changes in climate have caused nomadic barbarians to fall on their weaker neighbours and destroy the settled peoples and their irrigation systems. Provided the historical evidence is looked at from all angles and is checked against other physical or biological evidence, it can be a powerful tool in discovering more about changes in climate.

The well-documented and, in its second half, well-observed cold period of the Little Ice-Age was preceded by a few hundred years of warmer weather covering the time from AD 900 to 1300, the exact limits varying from one part of the world to another. This medieval warm epoch is sometimes known as the 'Little Optimum' and most parts of the world enjoyed climates as warm as any since the last ice-age. North America, Greenland, Europe and Russia all had two to three hundred years in which the average temperature was a degree centigrade or more higher than the temperature, say, in 1900. This may not sound a very exciting rise in temperature, but it was enough to allow cultivation 150 m. higher up the hillsides than today in England, and when the Little Optimum came to a rather abrupt end in 1310, vineyards in Germany had to be moved 200 m. downhill in order to survive. The traces of ploughing in the Durham and Northumberland fells went up to nearly 320 m. above sea-level, and it is reported that in 1234, complaints were heard because tilling the high-up fields left little room for grazing. This was a period of increasing prosperity in Scotland, which could be partly a delayed result of Anglo-Saxons seeking refuge from the Normans, but is more probably accounted for by the easier growing of food which was possible because of the warmer weather.

The Little Optimum was a time when barley, oats and rye were grown in the Viking colonies of Greenland. One rather macabre item of evidence for a warmer climate is the unearthing of the dead who were buried several feet deep in soil that is now permanently frozen. Viking expeditions seem to have travelled freely around the southern part of Greenland and as far to the north-east as Spitzbergen. These voyages contrast with the unsuccessful attempts from the sixteenth century onwards to force a north-west passage to the Pacific through the islands north of Canada. The difference could have been due to the temperature. This period of temperature increase made it possible for the Eskimo to spread as far north as Ellesmere Island, but when the Little Ice-Age brought cold weather they had to move to the southern coasts of Greenland and Alaska.

In North America, the Little Optimum brought more warm rain to the north-west and settlers moved into the valleys of Wisconsin and eastern Minnesota. However, changes in climate that are good for one may not be liked by others and the southern plains of the present-day USA suffered from drought. The tables were turned when the warm era switched rather abruptly to the early mutterings of the Little Ice-Age and while those in the north suffered from cold, the southern drier areas received much-needed moisture. A similar state of affairs exists today in the warmer southern parts of Europe, where a change to misery in the colder north probably means a more tolerable climate in the hottest months of southern Italy. These consequences of climatic change could have an important bearing on mankind if ever it was found possible to control the climate.

The times of the Little Ice-Age and the Little Optimum were by no means uniform in respect of temperature or rainfall. In the period between the two epochs there were good and bad spells of weather, and summers could sometimes be very hot even in a cold period. For example, the summers of 1433, 1434 and 1436 were warm or very warm; the summers of 1432, 1437, 1438 and 1439 were very wet, with floods and consequent crop failures; the winters were very cold and the springs wet and cold. (Those who think the weather is bad today should study some of the old records!) The first forty years of the sixteenth century were better years for Europe, but the glaciers were preparing to advance and a colder climate set in from 1550. This included severe winters and wet summers reminiscent of the periods following 1310 and the 1430s.

The last ice-age ended about 10,000 years BC and the historical period of climate in the civilized world has been fairly well recorded. The initial melting of the ice was accompanied by a fairly rapid rise in sea-level from 10,000 to 6000 BC, with another spurt between 5000 and 4000 BC, in what was the warmest period that has thus far followed the end of the glaciation. The total rise was 37 metres and the effects of the climatic fluctuations of the last few thousand years, described above, have had little effect on the sea-level. This indicates that the description 'Little' for the two episodes of warm and cold is correct. An approximate set of dates and temperatures (°C) taken from H. H. Lamb, *Climate: Present, Past and Future* is, for England and Wales:

Date	Epoch	Summer Temperature	Winter Temperature	Yearly Average Temperature
7000 BC	Pre-Boreal	16.3	3.2	9.3
4500 BC	Atlantic	17.8	5.2	10.7
2500 BC	Sub-Boreal	16.8	3.7	9.7
900–450 BC	Sub-Atlantic Onset	15.1	4.7	9.3
1150–1300	Little Optimum	16.3	4.2	10.2
1550–1700	Little Ice-Age	15.3	3.2	8.8
1900–50	20th Century warm decades	15.8	4.2	9.4

The divisions into the main climatic epochs have been made from botanical studies, both of pollens and of the associated vegetation. As the ice continued to melt around 8000 BC the temperature became higher than summers of today, and pine trees grew in southern England, with birch trees all over the country. The warmest times occurred from 6000 to 3000 BC in what has been called the Postglacial Climatic Optimum, and oak and beech trees appeared. In northern North America, temperatures were high, and Europe experienced the mildest winters of post-glacial times. It is not known whether it will be possible to go back to this warmest period without first passing through another ice-age but an examination of the results of the past always leaves some hope. The Sub-Boreal epoch, to about 1000 BC, was very warm at times, but fluctuations in temperature lowered the average given in the table. A significant summer cooling followed in the Sub-Atlantic period; the winters were mild and winds were strong. The 'Little' epochs have already been described, and the table indicates that at the present time we are in an interim stage with fairly mild winters and not very hot summers.

One of the interesting factors of the central period, from 1550 to 1700, of the Little Ice-Age is the increased variability in temperature from year to year and from one run of similar years to another. The frosts killed many oaks, elms and ash trees by splitting their trunks in what is called the 'great winter' of 1607–08. The sea froze on the east coast of Scotland so that one could walk out to the ships in the Firth of Forth. In 1634, a notably wet autumn and an early onset of winter occurred for the third year in a row, and frost and snow lasted for seven weeks in 1635. However, the autumn of 1635 was described as summerlike and extended until the end of November. An equally severe winter in 1683–84 with belts of ice along the Channel is reported to have killed much bird life and to have frozen wet ground to a depth of four feet, but it was followed by an early spring. Around the same period 1666, when the Great Fire of London took place, 1676, 1677 and 1701 were years with the hottest June or July to be recorded since regular observations began in 1659, and these years, in the worst period of the Little Ice-Age, rank as being among the 41 warmest months of the observing period up to today.

The Little Ice-Age epoch was also notable for increasing wetness of the ground, together with the spread of lakes and marshes in northern and central Europe. There is evidence of widespread rain and floods following the warm climate of the Little Optimum, together with high river-levels. As with the temperature, the rainfall also showed great fluctuations. In Spain, in the sixteenth and seventeenth centuries, agriculture was difficult on account of the spells of too dry or too rainy climates. Changes in North Africa seem to have occurred in step with the climate change in northern Europe. Cold periods in England correspond to droughts in the Sahara. It is possible that the whole pattern of northern hemisphere weather is pushed south by ice advancing from the Arctic. If we could only find enough circumstantial evidence to draw weather maps for past centuries similar to those we produce daily at the present time we should be able to begin to find the underlying mechanisms of the changes in wind and temperature patterns that have taken place in the past. Although the old records before the eighteenth century rely on deductions from many fragmentary and disconnected sources, it is hoped that the clues will one day fall into place and allow us to formulate a theory which will be valuable in looking ahead.

The small change of just over 1°C in average temperature between the Little Optimum and the Little Ice-Age had far-reaching effects on the population of the times when these climate changes occurred. The effect of climate on crops, on industry, and on health have always been, and still are, of vital interest to man although to a lesser extent nowadays on account of rapid rescue operations. Large moats had to be dug in the fourteenth century in Hertfordshire and Hampshire and buildings were moved upwards on the sides of valleys, in order to cope with the increased rainfall following the benevolent times of the twelfth century. At Moreton, near Bristol in Somerset, England, a whole village was shifted from the bottom of the valley, where it had no doubt been established to be near a small stream, to a drier site up the hill.

Some old bogs which had dried up and had become covered with trees returned to their earlier swampy state, and there are traces of wooden log trackways laid down to enable the people to make their way across the flat land between villages. It is probable that, all over the northern hemisphere, farms and villages were abandoned during the early wet, tempestuous days of the Little Ice-Age. In America, it was dryness in the rain-shadow of the Rocky Mountains that forced the more settled agricultural tribes to move to the south-east. In Europe, there was a tendency for migration to proceed in a southerly direction. In Europe, the sudden spell of bad weather that closed the benign Little Optimum episode of human history led to the first medieval movement from fixed abodes, and from 1315 to 1360 many villages were abandoned. A century later, between 1400 and 1480, another great exodus took place, apparently due to very cold winters followed by the characteristic variability of the summer rather than to the harvest-wrecking wet summers of the fourteenth-century famine periods. Long periods of very cold weather today are associated with steady anticyclones stationed to the west of the British Isles, which cause a clockwise wind circulation to bring freezing Arctic air down from the north. A neat piece of circumstantial evidence appears in the 1430s with the account of a Venetian ship which was blown down the English Channel into the Atlantic, where the ship foundered and the survivors drifted in an open boat clockwise round Scotland to Norway, presumably following the circuit round a stationary barometric high.

History books speak of the plague of the Middle Ages as the main destroyer of villages, and causing the abandonment of previously fertile farmlands. The notorious Black Death of 1348–50 certainly played havoc with the population of Europe, but when one looks into the evidence more closely it appears that the troubles that faced the agricultural population of those days began earlier in the century. It makes sense if the bad climate is given the part of the villain. The changes from warm, settled weather to rain and uncertainty took place early in the fourteenth century and reports from England, Denmark and the Dutch islands indicate that this is when the abandonment of earlier settlements began. The poor nourishment which resulted from bad crops and the cold wet climate, with dull springs and summers, must have had an enervating effect on the people just as much as it cut back production of the yield of wheat and the growth of animals. The drainage system was primitive in those times, and long winters with frozen ground probably upset the sewage disposal system, with consequent danger of the development and spread of disease. In fact, the general run-down condition of the country and the people is most likely to have been the precursor and the underlying cause of the period of plagues that beset Europe in the Middle Ages. Careful checking of the old tax records, which are an indicator of how the population was shifting, confirm this view of the relative parts played by plague and climate. Only ten per cent of eighty-four deserted villages in Oxfordshire and Northamptonshire are attributable to the Black Death. Rather it was the famine following those rained-out summers of 1310–20 that caused the population to decrease to a third from 1300 to 1327. Twenty years later, many of the villages were finally wiped out by the Black Death and it is this finality that probably produced the wrong impression in the history books.

Some places recovered from both the bad weather and the Black Death, but the argument in favour of climate being the underlying cause of movement of population is supported by the decay of more villages in the middle years of the next century. In John Rous of Warwick's history of England, fifty-eight villages in the Midlands are noted as being abandoned in his lifetime, which started in 1430 and extended to beyond 1485 when his history was written. As in the 1300s, climate research brings to light four or five rainy summers per decade in the 1430s and 1450s. In East Anglia, half of the 100 villages recorded in the Domesday Book of 1086 had gone by the time of the Black Death, probably because of change of climate. In some cases the increase in wind and generally unsettled weather turned light, fertile land into a desert of wind-blown sand. John Evelyn's diary in 1677 describes the area round Thetford as like the sands of Libya; this is in a place which in the earliest of days, back in the Stone Age, had always been light, easily cultivable soil. The origin of the Black Death is traditionally set in China in 1323, and this year followed a period of rain and flood which drowned about 7 million inhabitants and obviously left behind the usual trail of wrecked housing, lack of sanitation and human undernourishment on which plague and its carrier rats thrive.

It was not so much the wet weather as the decrease in warm days that caused the old Norse colony in Greenland to die out. The arrival of ice, which disrupted communications, helped the decline in this once flourishing settlement, and nearly caused Iceland itself to be evacuated during the worst parts of the Little Ice-Age. The

Eskimos seem to have moved in and killed what few settlers remained, probably because no ships could get in to provide supplies and trading goods. In 1350 one ship which called in reported, 'When we arrived at the place, no people were there, neither Christian nor heathen, only stray cattle and sheep of which we helped ourselves until the ships could hold no more.' However, if the sheep and cattle could live and breed it may be that the end was accelerated by human greed and avariciousness.

Efforts were made in the 1500s to resume the settlement, which was known only from the old maps, but there had been too long a cold spell for anything to be left of the old Norsemen. There may have been some intermingling of blood, because visiting ships reported that 'there are many houses where the householder does nothing but sell water which makes you quite mad. There they drink and yell and scream and fight and are without reason.'* This pub-like scene may recall the hard drinking of aquavit in the northern parts of Scandinavia.

Iceland experienced a pleasantly warm period from 1500 to 1550 and in 1540 it was possible to send an expedition to the old colony in Greenland, only to find that it had vanished. The population of Iceland was well over 70,000 in the early part of the fourteenth century; it dwindled to nearly half this by 1784, almost certainly because of the cold weather which probably had the effect of changing the economy from agriculture to almost a hundred per cent reliance on fishing. The population did not attain its warm-period maximum of AD 1000 until the beginning of the present century, since when the population has more than doubled. Iceland always had plenty of moisture and its principal crop is grass, on which livestock, mainly sheep, are grazed. Modern statistics show the effect of small differences in average temperature on the yield of hay from the grasslands: with a mean temperature of 7.6° C from April to October the yield was 4.33 metric tons per hectare, whereas in other years with a temperature of 6.8° C the yield was down to 3.22 tons per hectare. An overall picture of Iceland's climatic scene, which matches the population minimum in the 1700s, is provided by the fact that there were only 12 famine years in the 525-year span before 1500, while in the 204 years between 1600 and 1804 there were three times this number of famines. So it was the cold summers of Iceland and Scotland and northern Europe, the wet summers in the more temperate zone further south, and the droughts in North Africa which caused the population movements and the stagnation in numbers in the Little Ice-Age period. The year 1830 was a terrible one for the whaling industry. A thousand men were stranded on the ice of west Greenland having, however, secured a supply of food and aquavit, but the inexorable

* Erik Erngaard, *Greenland Then and Now*, 1972.

weather got them in the end. So these 'last of the Norsemen' are the relics of the Little Ice-Age, and looking back, we can see that the blooming period around 1100 was a transitory phase for those human beings who lived near what one might call a geographical-climatical boundary.

In some cases, there is evidence that man himself helped the changes set in train by the climate, and again a case can be made for asserting that it was fundamentally the change in climate that suggested to landowners a change in the operation and use of the land. The enclosure of land in England took place in the period 1300–1800, and the change of the Scottish highlands from small-holdings to grouse moors in the latter half of this cold epoch. Looking back on what must have been sad and troublesome times, it was probably inevitable that a new way of running the land would take place, since the system it replaced was only workable in the milder climate of the Little Optimum. On the other hand, it may be that enclosure of land and the farming of large units by powerful bosses, whether capitalist or communist, is a natural development that will always take place when human beings settle on areas of the Earth's surface.

There is no doubt that man has played a part in changing the landscape, by clearing forests to make room for cultivation and by conserving rainfall with irrigation ditches and dams. In some parts of the Middle East charcoal burning and goat and camel grazing have played a similar role in the removal of trees. The alteration of the Earth's surface may in turn have an effect on the weather, since the heat from the sun that is retained by the Earth is dependent on the reflecting and absorbing properties of the Earth's surface. The 'albedo' of a surface, the proportion of the light falling on that surface which is reflected back, is a convenient term for meteorologists when they are considering the contribution of the sun to the well-being of people on Earth. Since two-thirds of the Earth's surface is covered by the oceans, the behaviour of the sea surface will play an all-important part when we consider the total heat received from the sun. The heat that reaches the sea surface is rapidly mixed with the underlying layers of water and so the sea is a storer of the sun's heat. Solid earth, on the other hand, heats up quickly in the sunshine, and the heat is confined to a thin surface layer and radiated back again to the atmosphere when the sun sets. The sea keeps its heat, and by means of circulating currents like the Gulf Stream, it carries the sun's warmth all round the world. When the currents arrive in places where the sea surface is warmer than the air above it, heat is transferred upwards to the atmosphere. When the sea freezes, the sea behaves like land and the heat given to the air is much reduced. However, there is seldom much sunlight in those parts of the world where the sea freezes, so the changing

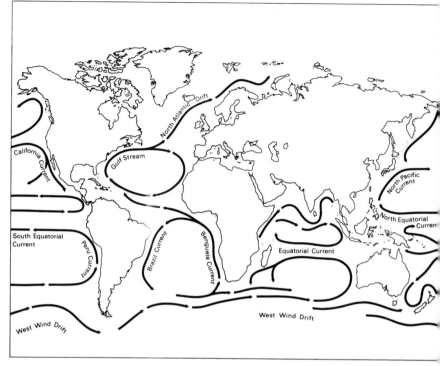

15 The main surface currents of the world

effect of ice formation on the total heat received from the sun and retained is very small.

It is the variations of the albedo on land, especially those parts of the land which are covered with snow and ice, that provide the big changes in heat received from the sun. Fresh snow reflects over 80 per cent of the sun's rays, while forest has an albedo of less than 10 per cent. Dry grass is 20–30 per cent, twice the value that is observed (8–20 per cent) when the grass is wet. Deserts give a figure of 24–30 per cent but rock is lower at 12–15 per cent. In general, the snow rejects the sun's heat while water, forests and wet ground absorb about 90 per cent or more of the heat available. The albedo of clouds varies from 20–70 per cent depending, as one might expect, on how dense the cloud cover is. Overall, the average of the Earth's albedo at the present time is 35–37½ per cent.

Changes in the heat given back to the atmosphere by the Earth will occur, then, when the amount of snow and its cover alters, and when the sea temperature changes. There is also an important effect produced by the evaporation of water associated with vegetation. The heat required to vaporize the water comes from the sun, and so there is a big difference in the heat transfer to the atmosphere in a given area if the ground is desert or is covered with grass or forest. This could produce an annual effect in places where the first rains transform a desert into a patch of green, or it could have a long-term effect when vast areas of forest are cut down in the interests of agriculture. Long-term changes in the climate of the Earth may depend on the amount of the sun's heat that is retained by the Earth and the atmosphere, and this amount will be mainly affected by the large thick ice-sheets of Antarctica and Greenland, both of which are very thick and require a large amount of heat to melt. The ice-sheets, being difficult to remove, provide a solid long-term area of high albedo and so ensure that a large part of the sun's heat that falls on the polar areas is reflected back again to space. The oceans may also have a long-term effect on the amount of sun's heat that is retained, since a large body of sea water, warmed by the sun, could conceivably circulate for centuries in the ocean depths before reappearing at the surface to give heat back to the atmosphere.

The heating of the atmosphere provides the energy for driving the wind circulation, and therefore the total heat reaching the Earth and the exchange of heat between the land, sea and atmosphere must play an important part in the world's climate. Clouds, which are dependent on the atmosphere and its circulation, themselves have an effect on the sun's heat budget to Earth. There is a great difficulty apparent here in deciding what it is that determines the overall effect on temperatures experienced at the Earth's surface and to what extent there are built-in stabilizing factors associated with the Earth and its atmosphere and its ice-caps. For example, if extra cloud shields the Earth from the sun's rays, it is probable that there will be less evaporation from the sea and therefore less cloud produced. Maybe, then, the long-term climate changes are dependent on changes in the supply of heat from the sun itself, and what human beings do to the Earth's surface is only of minor importance.

There is one satisfying facet of the study of past climate. The geological and botanical records laid down in rock deposits and fossil remains provide a record of what natural changes in climate were taking place before the world was sufficiently populated for man to have any appreciable effect on the landscape. When man did start to modify the evidence, at the same time he fortunately achieved the ability to leave written records of what the weather was like in his time. If he did not actually record temperature and rainfall and barometric pressure as he has done in the last three hundred years, he left enough clues to enable discerning climatologists to deduce what conditions were like.

To complete the climate picture back to the end of the last ice-age, some 12,000 years ago, it is necessary to rely on archaeological evidence rather than the historical picture that was so useful in studying the Little Optimum and the Little Ice-Age. At the same time, in view of the material that is available, which in turn fol-

lowed the movement of human activity, the scene changes from
England and north-west Europe and North America to the Middle
East and the eastern Mediterranean. Here there is evidence of agriculture and domestication of animals back to 9000 BC, in the warm
period following the last ice-age. During the next few thousand
years, civilizations developed in the sheltered valleys and flood
plains, where irrigation schemes could be constructed. Babylon
grew up in the long stretch of land dominated by the great Tigris
and Euphrates rivers. Numerous remains of old habitations show
how thickly the level tract to the east of the Tigris must have been
peopled. Today, this area, where Pliny recorded that local farmers
grew two crops of wheat each year, is largely a wilderness.

The Babylonians had good fertile soil, deposited by the large
rivers, which even today, together with the Karun to the east, are
building new land at the river mouths at the north end of the Gulf.
A few thousand years ago, Basrah was a port on the seashore; now
it is miles upriver, a solid inland city. It only needed a well-planned
and regulated network of canals to drain the primeval uninhabit-
able swamp and thenceforth to supply suitable summer irrigation,
for this part of the world to be a storehouse for a rapidly increasing
population. Babylon was a land of rich merchants and agriculturists, served for part of their energy budget by slaves. Today the
energy is supplied by indigenous oil, but the area is not nearly so
rich in agriculture, presumably because there has been a change
in climate. In the millennia before Christ, it is probable that warring
tribal neighbours wrecked the irrigation and drainage system, but
it is probable that this was merely a sharp finish to an inexorable
end of large fertility, caused by a steady decline in rainfall in the
area. Pointers to a decrease in water supply are provided by the
deserts of Sahara in North Africa, the Rajasthan area of India and
Pakistan and Sinkiang in central Asia. In all these cases, there is evidence of former animal and human occupation. Skeletons of elephants and rock drawings of wild animals are found in the Sahara;
vast ruins of buildings at Harappa in the Indus valley indicate a
prosperous area, through which Alexander the Great managed to
march his army; the Tarim basin of central Asia was crossed by
the caravans trading between China and the Roman empire.

It is probable that there have been fluctuations in rainfall in the
Middle East area rather than a regular decrease. It is possible that
increase in population denuded the countryside of vegetation and
altered the local climate, but it is more likely that the underlying
change is a downward trend in rainfall, coupled with the using up
of the underground supplies of water laid down in the wet period
of the last ice-age. In England, the earliest settlements around 4000
BC were of Neolithic people who grew crops and kept sheep. This
was in the warm period following the ice-age, and there was a tendency to live on the tops of the chalk hills. As the climate became

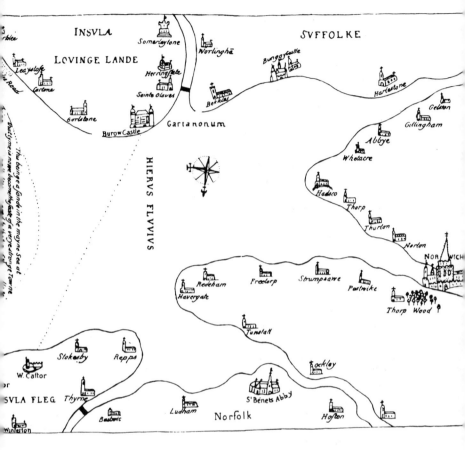

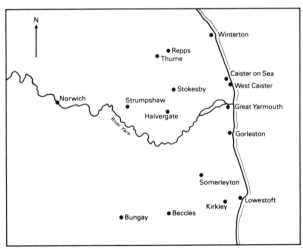

6/17 This map (*above*), taken from a fourteenth-century MS, shows what the East Anglian coastline looked like about AD 1000 when the sea-level was higher (south is at the top). Comparison with a modern map shows that the city of Norwich was then a coastal town on a broad estuary, while Great Yarmouth was represented by a submerged sandbank drier, there was a tendency to move into the valleys but, as we have noticed earlier, when the cold wet weather started after 900 B C, the early Iron Age people drifted uphill once more. In the centuries leading up to the birth of Christ, the climate in northern Europe got steadily worse, and what evidence remains suggests conditions reminiscent of the Little Ice-Age of the late Middle Ages. It was cold even as far south as Italy, and the Roman river Tiber froze over four times during the period 400 to 150 BC.

One of the ways in which ancient climates can be reconstructed is by finding out what animals lived in various places, and one way of learning about the animals is from old drawings and paintings. For example, in the warm days of 4000 BC there were elephants, ostriches, lions and giraffes in the Nile valley, and these were all shown in rock drawings and prehistoric rock paintings. These animals, together with others such as the hippopotamus, which is shown in one fresco being hunted from canoes, have moved to wetter lands to the south and one must infer that the Nile area became drier than it was 7,000 years ago.

Paintings on canvas in the sixteenth century can be used to support the climatic reconstruction of the Little Ice-Age. This is especially true of some of the works of Pieter Breughel the Elder, which concentrate on the landscape rather than on people as was generally the case with previous painters. The Dutch and Flemish landscape pictures show severe winters, but they also allow some observations to be made of the cloudiness that prevailed at the time of painting during the summer months.

Stone Age art in Scandinavia shows bears, elk and whales, and hunters using skis and boats. It was much more pleasant hunting in those warmer days following the ice-age, but even during the ice-age reindeer, bison and the great mammoth were hunted in France on what were then often grassy steppes and tundra bordering the ice-cap. These ice-age activities were recorded for posterity in the famous wall paintings of Lascaux and other caves in France.

5 Climatological clocks

The earth is a wonderful storehouse of information for those who have the eye and the imagination to look at what exists in the air and the rocks and the oceans. The museum of natural phenomena allows us not only to put ourselves in the place of the ancient French cave-dwellers who knew the mammoth, but also to re-enact some of the geography of the days of the ice-ages. There are geophysical, physical and geological methods now available by which it is possible to look into the past and compare the temperature of north Europe with that of different parts of the Earth today. About 12,000 years ago, prehistoric man witnessed the break-up of the last of a series of ice-ages which had occurred during the past 125,000 years. Perhaps one should say latest, for, as will be discussed when all the evidence has been assimilated, there is a distinct possibility that further ice-ages lie ahead. One of the most valuable outcomes of the work of the modern school of climatologists is to improve our knowledge of terrestrial phenomena so that due warning may be given to future generations of impending climatic changes, both the smaller variation in climate that affects agriculture and the long-term severe descents into the true glacial periods.

The ice-ages experienced in the northern hemisphere are periods in which there is a marked lowering of temperature, accompanied by glacial conditions not unlike those which now characterize the polar regions. Great ice-sheets flow over the land, wearing away the rock and depositing the debris that is collected en route, in just the same way as glaciers flow inexorably down Alpine valleys and produce a moraine of rubble at the bottom end of the glacier. Geography students will remember that the grinding rivers of ice gouge out a gentle U-shaped path in the rock over which they are travelling, as opposed to rivers, which have little sideways spread and cut a sharp V-shaped valley. Although ice appears to be hard and solid when struck with a pick, it has in slow time the properties of a fluid, and therefore under the forces of gravity it flows very gradually downhill.

The moving ice scratches the old hard rock on which it rides, and these signs, together with the debris such as boulder clay and the evidence of erosion, can be seen today where glaciers have retreated in icy mountain regions. Boulder clay is the result of the grinding and scraping by the ice of the rocks over which the glacier travels, and therefore varies both in chemical composition and colour. It will be red when the ice has moved over Old Red Sandstone, and nearly white when chalk was the underlying rock. The boulders vary in size from pellets to masses many tons in weight, and they show signs of abrasion and, as with the clay component, they also provide evidence of the track of the ice. Some old geological theories ascribed boulder clay to the debris dropped from icebergs in periods when the land had been submerged. However, boulder clay is now one of the foremost indicators of old glacier movement and therefore of previous cold climatic eras.

The distribution of boulder clay and of abraded rock allows reconstructions to be made of the appearance of the land at such times as 18,000 years ago, when the last ice-age was at its most far-reaching extent. The Arctic ice-cap had formed and had spread southwards to cover Canada completely and the U S A well beyond the Great Lakes. All of Britain and northern France were in the grip of the ice, which was up to 10,000 feet thick in the Baltic area, and more than 12,000 feet thick over the island of Novaya Zemlya, while over Hudson's Bay the surface of the ice was more than 7,000 feet above sea-level. The weight of ice pushed the land down, just as a heavy deck cargo causes a ship to sink deeper into the water. In general, the continents behave as if they are floating in the rock layers of the mantle 50 miles below the surface and one of the delayed signs of the ice-age is the gradual recovery of land after the ice-sheet has melted. Careful measurements show that Scandinavia is even now rising to gain its true equilibrium level. In Greenland, and the Antarctic continent, similar thicknesses of ice-sheets have been measured by expeditions during the past forty years, and although we are not in the throes of an ice-age, we can gain a clear impression of what the face of Europe was like.

At the foot of the ice-sheet, which ended with massive glaciers fingering out to the north of Germany, there was a plain of permafrost. This was similar to the deeply frozen ground in the North Slope oilfield area of northern Alaska, and this in turn gave place to a stretch of marshy tundra like that in Russia, where the old cave-dwellers hunted the wild animals they so artistically recorded on the stone walls. The warmer climates farther south allowed forests to grow, as they do today except that the forests are now largely replaced by agricultural land. The forest/agricultural belt was much smaller 18,000 years ago than it is today, and accounts for most of the land taken up by the belt of ice spreading from the north. The permafrost and tundra covered the same extent of about 600 miles although displaced some 1,200 miles south during the ice-age. The Mediterranean climate, instead of covering the Mediterranean Sea as it does now, was pushed down to North Africa and encroached on the desert, which was some 400 miles narrower than it is today. The rest of the picture, wild-game savannah country and tropical rain-forest, was much the same as it is now. Ice-ages then, although providing an ill wind for those who

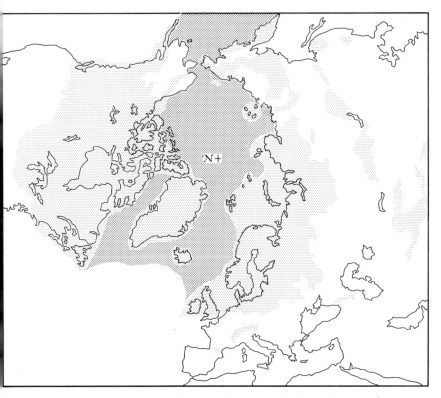

18 The last ice-age. The shaded area shows the furthest advance of the ice – in Britain as far south as the Thames, in the U S A down to the latitude of Detroit

lived in temperate climates like Britain and the northern U S A, were a boon to the desert-dwellers of the Sahara, who received benign, rain-laden breezes in place of the scorching desert sun. For anyone who considered planning a long way ahead, some property in the northern part of a northern hemisphere desert would be a sensible investment. However, we must learn more about the mechanisms of climate change before any accurate forecasts can be made. Our knowledge at the present time indicates tens of thousands of years between ice-ages, so we have a long time to go.

The temperature difference between the ice-age and our present-day interglacial period is about 10° C, much greater than the 1° C difference between the Little Ice-Age and the Little Optimum described in the last chapter. The 10° C is a yearly average and the fluctuations around the average were very much greater than this. On the other hand, similar large fluctuations from year to year do occur at the present time, so that an ice-age will bring nothing new to man, and nothing that he cannot experience by going up to the far north today. The last ice-age lasted about 60,000 years and was preceded by a pleasant warm spell which reached its height about 120,000 years ago. The sea then was 50 feet higher than it is now, and it was at this time that England was separated from Europe for the first time. Five thousand years later more ice formed and withdrew water from the sea, and during the farthest advance of the Arctic ice-sheet so much water was locked up as ice that the sea-level dropped to 30 feet lower than it is today. This meant that the land area increased towards the edge of the continental shelf, that fringe of shallow water around the continents which is geologically part of the land and on which we are finding oil. Extending the land in North America or north-west Europe was of no great advantage to man, since the whole territory was in any case covered with ice, but since sea-level is the same all over the world, the lowering of sea-level caused significant changes in the subtropical regions of the southern hemisphere. The southern part of South America widened out considerably, and many land connections must have appeared in the shallow area of South-East Asia between Australia and Indonesia.

There is still plenty of ice held in the Antarctic and Greenland ice-caps and its thickness is known to be such that, if it melted, the sea would rise by about 300 feet, thus reaching the level of the warm climate days over two million years ago before the ice-age period of the last million years started. A rise of 300 feet would drown most of the large cities of the world, which have been placed more or less at sea-level since many of them started as ports, or places where river estuaries could be crossed.

After the last hot climate of 120,000 years ago, the writing was on the wall with regard to ice-ages. It is easy to say this while looking at the graph of temperature for the last 150,000 years, but for anyone at the time it would have been even less possible than it is today to make any kind of reasonable forecast. There was a recovery from the cold period of 115,000 years ago, but this warm period of 10,000 years or so was again followed by an average temperature drop of a few degrees to give permanent snow cover to a large part of North America and Europe. A quick recovery over a period of 2,000 years led to 10,000 to 20,000 years of reasonable warmish climate, but this was the precursor to the proper ice-age when, 70,000 years ago, the average temperature dropped by about 5° C and the ice-cap began to grow. Although there are fluctuations in the yearly average temperature, the record is firm on the fact that in north-west Europe and North America the average temperature from about 70,000 years ago until 10,000 years ago was some 10° C less than it is today, and these temperature records agree with the geological evidence for the advance of the ice-sheet from its present position way up in Greenland and the Canadian northern islands down to the Great Lakes and the south of England. After 50,000 years of icy misery, a final extra cold spell occurred some

20,000 years ago, after which the temperature climbed somewhat unsteadily to the warmest part of our post-glacial world history some 10,000 years ago, as described in Chapter 4.

The ice-sheet appears to have reached its greatest extent between 17,000 and 25,000 years ago, but in the 60,000 years of the ice-age, the climate in France, southern Britain and south-eastern Canada would warm up for periods of a few centuries or more, allowing forests of spruce and pine to grow, and to leave their traces for the climatological detectives of today to use as clues. With ice to the north and east, western Europe, including London and parts of the Netherlands, experienced a much more temperate climate than one might expect from the air circulation patterns that we observe in the twentieth century. It is apparent that, in addition to explaining the main average temperature drop of the 60,000-year-long ice-age, climatologists have to account for the hundred-to thousand-year excursions into warmer conditions, and, looking at present-day weather, also find mechanisms for the rapid variation that occurs from year to year, giving freak snow-bound winters and rainless summers. There must be many different effects at work, and it is possible that one cause enhances or even triggers off an unstable situation produced by a completely unassociated set of circumstances. Such complications not only make it difficult to unravel the mystery of variation and change of climate, but they also decrease the credibility of any forward warnings which might be of value to mankind.

The remains of vegetation provide the evidence for warm spells during the ice-age, and they are backed up by other specific measurements that can be made by physicists, and by geologists examining the detail of rock deposits. Sometimes tree stumps or leaves of trees are preserved, usually in waterlogged conditions such as muddy lakes or peat bogs. The position of the tree line is a good indicator of climate, and tracking the ancient spread of forests can give an insight into what conditions were like in the past. More often microscopic spores from trees provide the best evidence of what the vegetation was like. These spores are very resistant to decay and can be traced back 400 million years; since one branch of an oak tree yields about 100 million spores, there is a good chance of some surviving. The snag about pollen analysis is that the pollen can travel great distances on the wind and so the distribution on land can be deceptive. However, by careful study of different species and different sizes of pollens, which settle at different rates and so have distinguished paths, it is possible for the expert palynologist not only to plot the vegetation limits but also to deduce something about the past wind circulation. A 14–16° C summer temperature is needed for spruce, pine and birch to thrive, so that the tree evidence provides a measure of temperature in the past, and thus extends back into geological time both the direct temperature measurements of the last few centuries and the temperatures deduced from archaeological observations.

Trees can provide a direct measure of age going back thousands of years, so it is possible to date the climatic evidence that the trees themselves provide by their old tree-line positions. When a tree grows in a climate where there are distinct summer and winter seasons, clearly marked growth rings are produced, one ring per year. Since some of the giant sequoias and similar coniferous trees live for thousands of years, counting tree-rings provides an accurate dating method for considerably farther back than even archaeology takes us. We can extend the record of present-day trees by counting the rings of fallen trees and matching up the overlapping rings. This is possible because, owing to variations in rainfall and temperature from year to year, the rings are not exactly equal in width and a recognizable pattern is produced. The bristlecone pine of the north-west U S A is probably the longest-living species of tree, some specimens being 4,000 years old. This has allowed tree-ring experts to produce a reliable calendar back to 7,000 years ago, and by examining the detail of the rings it is possible to make an assessment of climate over this period. Trees in the uniform climate of the tropics do not show annual rings, but exhibit thin dark lines when dry spells of weather occur. The best trees for counting the years are old hardwoods which have grown on the outer parts of the forest so that they are exposed clearly to the seasonal changes in weather.

There is no reason why the tree time-scale should not be extended if enough old preserved trees can be unearthed to give an overlapping record. The year rings are responsive not only to variation of exposure to the elements, but also to attack by insects; they are affected by summer temperature and lack of moisture, and sometimes a ring is missing on account of one of these causes. However, by using several trees for comparison, it is usually possible to get a true count, and one research group have checked their results by repetition and found them accurate to plus or minus one year as far back as 3525 BC. It has been suggested that tree-ring growth can be correlated with barometric pressure, thus allowing the type of weather map produced by forecasters today to be drawn for days of hundreds of years ago. The long-lived bristlecone pine is an indicator of early August–September frosts, which have occurred in historical times when large volcanic activity was reported, and therefore are probably a reflection of temporary loss of sunshine due to volcanic dust in the atmosphere. The oldest frost indication noted by this particular investigation is 1626 BC, and it is a pleasant thought to associate this with the gigantic volcanic blow-up which destroyed the Minoan civilization on ancient Crete, and which historical records ascribe rather vaguely to within a few centuries of this date.

The precise counting of tree-rings has provided the best check on what is one of the most powerful age-dating tools invented by modern physicists. There is a form of carbon which is radioactive, and the radioactive atoms throw out detectable particles when they change to stable, inert nitrogen atoms. The rate of break-up of the radioactive carbon (called carbon fourteen, or C-14, because its atom has 14 times the mass of an atom of hydrogen instead of the normal carbon atom's 12 times) is very regular, so that one-half of the C-14 decays every 5,730 years. If we start with a gram of C-14, half a gram will be left in 5,730 years, a quarter of a gram after 11,460 years, an eighth of a gram after 17,190 years and so on, so that by measuring how much C-14 there is left in a sample whose original size was known, the age of the sample can be calculated. The amount of C-14 can be measured very accurately since the particles it gives off when it changes to nitrogen can be counted, and the best experimenters can measure one five-thousandth of the original sample to provide ages of 70,000 years, which is a very useful span backwards when considering past world climates. Now, while we can measure the amount of C-14 in a sample, say, of wood, which is nearly all carbon, we need to know how much C-14 there was when the wood was formed, if we are to calculate the age of the trees from which the wood was derived. The supply of radioactive C-14 is provided by bombardment of nitrogen atoms in the atmosphere by cosmic rays from outer space. The newly formed radioactive carbon atoms soon join up with oxygen to produce carbon dioxide, which in turn is used by plant and animal life as their source of carbon. When the plant dies, there is no replenishment of radioactive carbon through the CO_2 of the atmosphere, and the C-14 content starts to decrease, halving its concentration every 5,730 years, and providing a carbon clock, which as we have seen, can in expert hands tell the time in years back to 70,000 BC.

There always seem to be snags in new discoveries, and in the case of the C-14 clock, it appears that the atmospheric bombardment of nitrogen has not always been taking place at exactly the same rate as today, probably because of changes in the magnetic field and the solar wind around the Earth, which affect the path of incoming radiation (see p. 86). This is where the tree-ring count comes to the aid of the physicist by giving him a series of standard samples of known age back to 7,000 years ago. This biological type of calibration can be extended by further help from geological observations. Some rocks formed of mud and clay sediments show a neat pattern of layers which have been laid down each year when rivers in flood bring their annual contribution of debris from upstream and deposit it in a lake or on a seashore. Eventually the sediment dries up and is covered with other deposits to form a rock which the geologist can examine. The 'varves', as the layers in the sediment are called, are the geological counterparts of the tree-rings and can be counted and overlapped and matched as was the evidence from different trees. The measurement of varves is useful in its own right for determining the rate of deposition of sediments, and marking the time when retreating ice-sheets left small lakes behind them, but it also can extend the tree-ring calibration of the C-14 method for a few thousand years. Like the tree-rings, varve counts can give relative dates at various times back in geological history, even if a continuous overlapping series is not available to take a direct count up to the present day. The variations in thickness of the fossilized mud layers could also provide another clue in the detection of previous climates, since a year of heavy rainfall will normally produce a thicker deposit than a dry year. A rather devious kind of clue is provided by an examination of what material each year's individual layer contains. An example from the coast of California is an abundance of fish scales, which indicates fluctuation in the fish population, which in turn suggests changes in the ocean currents.

There are two man-made sources of uncertainty in the C-14 dating measurements. In the first place, the burning of coal and oil has produced carbon dioxide which is not radioactive, and this has diluted the C-14 in the atmosphere. Fortunately, the sea absorbs a good deal of the carbon dioxide, so that the man-made CO_2 causes only a few per cent dilution of the active C-14. The letting off of nuclear explosions has the reverse effect, since radioactive C-14 is produced by the atomic bombs, thus counteracting the contribution of the CO_2 from burning fossil fuels. These problems are avoided by the use of standards of known age and by checking the amount of radioactive C-14 in the atmosphere, but care must be taken in using the C-14 dating method, since variations do occur, for example from the north to the south hemisphere, on account of local addition of 'old' carbon dioxide from melting ice or from the deep waters of the oceans.

Carbon is not the only element which has atoms of different mass, or isotopes, which can help in tracking down past climates. The common element oxygen is normally present as oxygen sixteen (O-16), but it has a useful isotope O-18 (as well as an O-17). In 1947 Professor Urey suggested that the ratio of the amount of O-18 to O-16 in calcium carbonate deposited on the sea-bed might prove to be an indicator of the temperature at which the marine animal lived which had manufactured the calcium carbonate for its shell. When the shell is formed, the concentration of O-18 is increased because the calcium carbonate with O-18 in it is heavier than that with O-16 and so it deposits more rapidly, especially if the temperature is low. A rise in water temperature makes the difference between O-18 and O-16 in calcium carbonate and water decrease, so that theoretically a measurement of O-18/O-16 in calcium carbonate, which is a prime constituent of the shells of marine organisms, provides a measure of the water temperature that existed when the shells were made.

As always seems to happen when a great new method of probing into the climatic past is invented, there is a problem with the oxygen method of measuring sea-water temperature. The temperature results of the O-18/O-16 ratios in shelly remains of sea-life depend on the water composition being constant, and it is just at the time when we are particularly interested in water temperature that it is probable that the ratio in the water itself changes. When the great ice-sheets form during the severe periods of ice-ages, a great deal of water is locked up in the ice, and this water comes from the evaporation of the sea to form the snow which falls and forms the basis of glaciers and ice-sheets. Since water with O-16 is lighter than water with O-18, it evaporates more readily, which means that ice-sheets are low in O-18 compared with sea water. When the ice-sheets melt, the O-18 of the sea is diluted, and when the ice-sheets are at their maximum, the O-18 in the sea water is at its highest concentration. The temperatures derived from the oxygen-isotope ratio must then be corrected for the amount of sea water locked up in the ice-caps. This may appear to be a disastrous snag to the temperature measurements, but, provided we understand what is happening, the measurements can be used to assess the extent of the world's ice at any past age, rather than the actual temperature of the surface water. If we measure the O-18/O-16 ratios in selected materials, we can, as will be explained later, find out several different properties of the ice-ages. The determinations of the O-18/O-16 ratio are made with a mass spectrograph, which today is a familiar tool in most geological laboratories.

It is possible to make a check on the amount of water held as ice in the Arctic and Antarctic ice-sheets, because, as noted previously, the removal of sea water to form huge masses of ice causes a change in sea-level. The sea-level at various eras of the geological past can be detected by the remains of old beaches and flat plateaux formed by wave action at the fringes of the land. The existence of ancient coral platforms also provides a clue to old sea-levels, since coral reefs only grow in shallow water. Incidentally, since coral animals only thrive in warm water, their existence also tells something about past climate. It is important to look at old sea-level evidence in places remote from the polar regions, because the ice-caps themselves depress the land and so give false levels for old beaches and wave-eroded platforms.

Before the idea of using oxygen isotopes to measure past water temperatures was proposed, a great deal of climatic history over the past million years was being unravelled by examination of the fossil remains of marine life in cores of sediment collected from the floor of the oceans. Here the rule is, 'big fish eat little fish and little fish eat lesser fish and so on.' The plankton, which are the animals and plants at the beginning of this food chain, leave as their testament to their short existence the most beautiful microscopic fossil remains. Some species live in cold water and some prefer it warm, so that careful separation of the fossils of these 'foraminifera' tells the geologist a great deal about life in past times as shown by the layers in the sea-bed core sample. Many of the foraminifera build their shells in a spiral, and some of them have a most peculiar idiosyncrasy. A spiral can form either with a clockwise or with an anticlockwise twist, and for some reason, at present best known to themselves, a few species of spiral-shell animals coil to the left in cold Arctic waters and to the right when in warmer water. One particular friend of the geologist, *Globigerina pachyderma*, makes its change-over at the particular temperature of $7.2°$ C, and therefore can be used as another type of geological thermometer.

Great care has to be taken in examining cores taken from the sea-floor, since the steady deposition of fine sediment interspersed with fossils of animals is often interrupted by the slumping of large blocks of sediment down the slopes of the uneven parts of the ocean, and by turbidity currents, which are fast-moving streams of mud-laden water which roar down the continental slopes like avalanches in the mountains, leaving behind a confused record of sediment which completely upsets the steady rain of material which is needed for an accurate reading of what occurred in the past. Another trap for the unwary investigator is provided by the fact that calcium carbonate fossil remains dissolve in very deep water and so the pages of the geological history book are destroyed. However, by selecting cores which are on rises in the sea-bed remote from the possibility of sideways mud movement, and which are regular in texture, and for which the upper part can be checked for age by carbon-14 measurements, it is possible to collect laboratory material which can yield a true story of past history.

The selection of the species of fossil to be examined depends on what information is required from the sea-bed core sample. It is necessary to look at the life history of similar species living today. For example, some marine life stays near the surface of the sea because it needs light in order to live. The radiolaria, with skeletons of silicon which are resistant to solution in the deepest waters, provide an indicator of the sea-surface temperature, because their abundance can be checked against samples of similar animals collected today. Some of the foraminifera, with the beautiful calcium carbonate shells, do not provide a proper indication of sea-surface temperature because, although they start life in the sunlit surface waters, they move deeper to thicken their shells in their old age, and therefore reflect the temperature and oxygen-isotope composition of the oceans in general. Many more details of the fossil remains and the abundance of various species make it possible to check one core against another and to note where rates of sediment changes occur, so that a reliable calendar of past geological events can be established.

The geophysicist helps in the correlation process by measuring the magnetic field of segments of the sea-bed cores. Measurements of the magnetic field of rocks, such as are formed by frequent (in a time-scale of millions of years) volcanic outpourings in actively volcanic places such as Iceland, show that the Earth's magnetic field, although north/south at the present time, has often in the past reversed, so that a compass needle would point south instead of north. These reversals have been carefully measured on the geological time-scale and in themselves form another dating method back into the past, which is used to check the time-scale of the sea-bed cores. The reversals of the Earth's magnetic field upset the stream of particles, such as cosmic rays, which hit the Earth from outer space, and this may have a noticeable effect on the mutations of some of the species of minute animals found in sea-bed cores. There is evidence that some species came to an abrupt end at the time of magnetic reversals. The extinction of animal species of any kind, large or small, could be an indicator of climatic change, but before the climate is blamed for any particular happening of the past, it is important first to rule out all other possible causes of disaster. In some cases, such as a coincidence between extinction of animals and exceptional volcanic activity, it may be that there was an accompanying disturbance of the climate which was directly responsible for the animal behaviour.

The oxygen-isotope ratio measurements can be applied to cores from the Arctic and Antarctic ice-sheets. The water supply for the ice and snow is mostly derived from warm tropical water evaporation, and therefore the oxygen-18 content of the ice provides a measure of what change in temperature occurred when this warm moisture was deposited. Measurements on samples taken from Greenland indicate that 15,000 to 70,000 years ago temperatures were 20° C colder than now, but there is a qualification which reduces this value to half, namely the fact that the land stood thousands of feet higher than today. This only goes to show that in these searching investigations into the past climate, one cannot be too careful.

An idea of the temperature prevailing at the bottom of the oceans can be found by the oxygen-isotope method, if suitable fossils of animals that live at the sea-bed can be found. However, the results in this case are even more difficult to interpret than are those for the surface waters, because of unknown effects of the sinking of cold heavy water in the polar regions and the general slow circulation of the whole of the water of the oceans. On the other hand, all these observations must be fitted into the whole puzzle if the truth of what has occurred on this Earth is to be discovered. It is probable that there was an overall drop in temperature of the oceans of a few degrees centigrade during a large part of the ice-ages.

While the microscopic plankton were responding to changes of water temperature by increasing their abundance or changing their direction of coiling, land-based insects were also reacting to the prevailing climate. A beetle which eats caterpillars and inhabits oak trees, for example, cannot live if the mean July temperature is below 15° C, while the house-destroying longhorn beetle likes it above 16.5° C, and therefore did great damage in late Georgian times in London, but died out until the spell of warm summers from 1934 to 1953, when they re-emerged from the woodwork. Moths and butterflies have their particular temperature tolerances, while some of the mildews and other tree diseases thrive on wet summers. A present-day pest, the beetle that has spread the Dutch elm disease, likes mild winters and it is hoped that a series of cold years may stop its ravages. For warmer climates locusts, mosquitoes and fleas all have their climatic likes and dislikes, and since there are about a million different species of insect in the world, careful observation of old insect remains makes it possible to check on temperatures of past ages since the hard parts of the insects are preserved. Beetles are among the more useful species found in old lake deposits, and their fragments can be identified in much the same way as the pollen remains of plants and trees. Sometimes the 'beetle temperature' changes more rapidly than that estimated from the spread of vegetation, but this is what might be expected since insects are much more mobile than plants and trees. The insect evidence shows that for about a thousand years during the last ice-age, round about 42,000 years ago, the summer temperature was on average warmer than it is today, although the good climate was not long-lasting enough to allow trees to spread back to northern Europe from the Mediterranean region.

Opposite: a glacial valley in Merionethshire, Wales. The U-shape is a characteristic result of the passage of ice

Above: the Siberian tundra is a permanent storehouse of evidence for ice-ages thousands of years ago: this baby mammoth was discovered in 1977, after lying in frozen ground for about twelve thousand years

Below: during the 'Little Ice-Age' (1550–1700), the River Thames froze over several times, as recorded here in a detail from a painting by Abraham Hondius, *Frost Fair on the Thames* (1683–84)

Opposite: America's Grand Canyon, with its many rock strata, makes it possible to turn back the pages of geological and climatic history layer by layer and period by period

A nineteenth-century engraving of the glacier of Argentière and a photograph of the same site, taken a hundred years later in 1966, record the movement of the glacier and the alteration in the tree-line

Desert areas such as the Calanscio Sand Sea, Libya, had a greater water-supply in the millennia before Christ

If the ice-cap of Gunnbjørns Fjeld in Greenland were to melt because of a drastic heating-up of climate, the oceans of the world would rise by about 6 m (20 ft)

An artist's reconstruction of the moment, tens of thousands of years ago, when the ice-sheet threatened the mammoth with extinction

Typical glaciated valleys: the Brecon Beacons, Wales

Llanberis Pass, Wales, was carved out by the slow and heavy movement of a glacier

Much of the Middle East, like Tamrit in Saudi Arabia, is permanent evidence of a decrease in rainfall coupled with a depletion of underground water supplies laid down in the wet period of the last ice-age

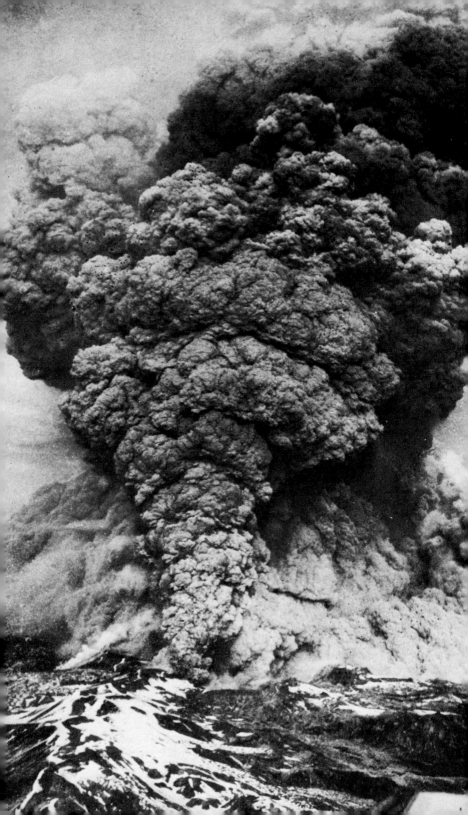

Opposite: volcanoes like Phyehue in Chile affect the climate of the whole world by the vast amount of dust particles they emit into the atmosphere

Above: whirlwinds and scant rainfall in this Utah desert combine to release great amounts of dust into the atmosphere, affecting weather conditions for miles around and prolonging the basic arid nature of the terrain

Below: the customarily hot climate of Dallas, Texas, occasionally sends up into the atmosphere great quantities of warm, moist air whose spiral rise accelerates into the devastating tornado, a funnel whirling at more than 600 km per hour (375 mph)

Opposite above: a meteorologist's interpretation of the satellite photograph (*opposite below*) which was taken on 12 January 1979 from Geostationary Operational Environmental Satellite 3 and illustrates many of the important features of weather. The sun glint indicates an area of calm seas. Meteorologists use successive pictures to track the movement of cloud systems

Views of North Africa and western Europe taken by the Nimbus 1 weather satellite. Each of the strips of three frames covers an area of approximately 1,295,000 sq. km (500,000 sq. m)

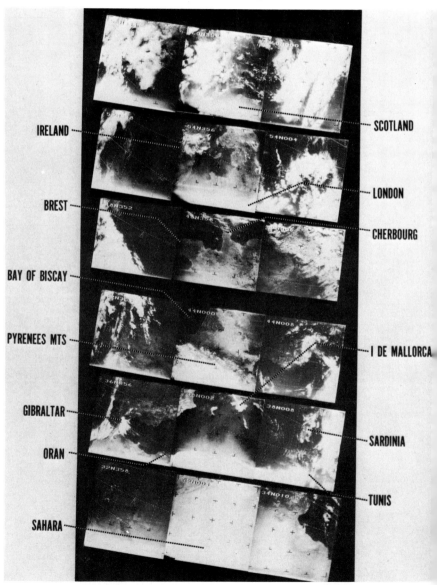

IRELAND — SCOTLAND

LONDON

BREST — CHERBOURG

BAY OF BISCAY —

PYRENEES MTS — I DE MALLORCA

GIBRALTAR — SARDINIA

ORAN —

TUNIS

SAHARA —

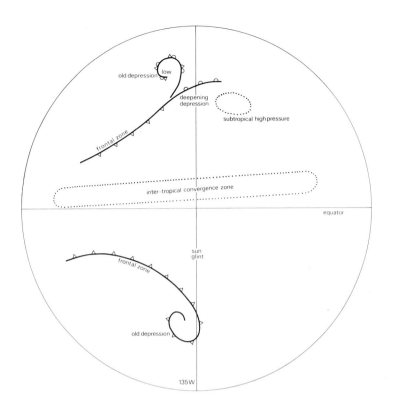

old depression   low

deepening
depression

subtropical high pressure

frontal zone

inter-tropical convergence zone

equator

sun
glint

frontal zone

old depression

135 W

Opposite above:
From 6,100 m
(20,000 ft) Hurricane
Esther looks like a
tightening spring,
preparing to release
its tension and
unleash destruction

Opposite below:
The eye of
a hurricane,
photographed in
1961 from the
Mercury-Atlas 4
spacecraft

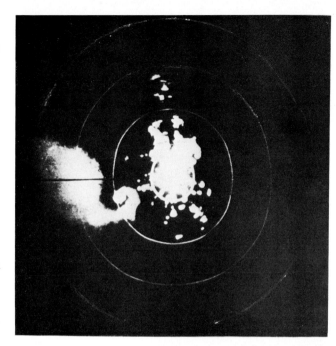

A radar photograph
of a cyclone, an
intense depression
or trough, capable
of great destruction
in tropical areas

Twin hurricanes
Ione (left) and
Kirsten present
swirling patterns
as they move across
the Pacific Ocean
between Hawaii and
Mexico in this
picture taken by a
satellite in August
1974. This and other
American NOAA
(National Oceanic
and Atmospheric
Administration)
satellites are helping
in the prediction of
tropical storms. Of
particular interest
are the well-defined
central storm 'eyes'
and spiralling
shadows. The polar-
orbiting spacecraft,
launched in
November 1973,
photographs the
entire Earth each
day in both visible
light and infra-red

Cloud seeding over the USA. In the future it may prove possible
to temper the force of hurricanes by causing the water vapour at the
centre of a cyclonic disturbance to drop prematurely as rain before it
condenses, giving off the heat that will power a full-grown hurricane

The force of this storm, photographed 15,250 m (50,000 ft) above
western Oklahoma, was estimated to have been fifty times greater
than that of the atomic bomb which struck Hiroshima

The immense energy of a hurricane is released and converted into waves battering a seafront promenade

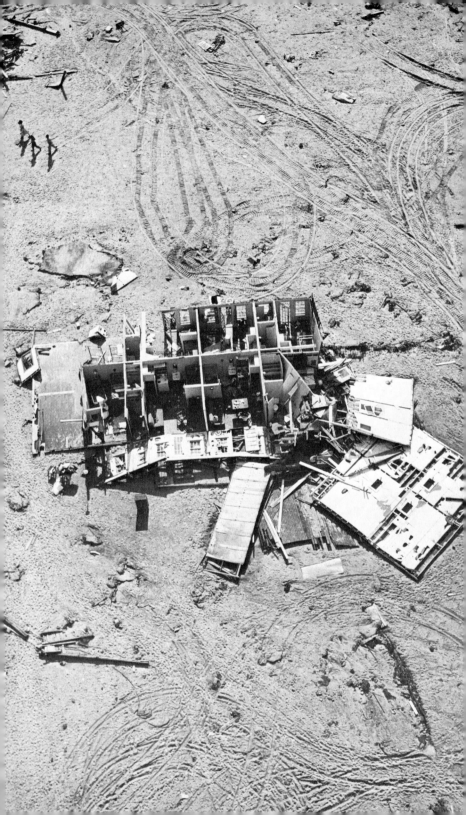

	North America central sector	Britain	North European plain	European Russia	Alps	Emiliani's Stages
Postglacial		Flandrian	Flandrian			I
Upper	Wisconsin	Devensian	Weichsel	Valdai	Würm	2
	Sangamon	Ipswich	Eem	Mikulino	Riss-Würm	3
	Illinois	Wolstonian (or Gipping)	Saale	Moskva	Riss	4
				Odintsovo		5
				Dniepr		6
Middle	Yarmouth	Hoxne	Holstein	Likhvin	Mindel-Riss	7
	Kansan	Anglian (Lowestoft)	Elster		Mindel	8
	Afton	Cromer	Cromer	Morozov	Günz-Mindel	9
		Beeston				10
		Paston				11
Lower	Nebraskan	Baventian	Menap	Odessa	Günz	12
		Antian	Waal		Donau-Günz	13
		Thurne	Eburon		Donau	14
		Ludham	Tegel (Tiglian)		Biber-Donau	15
		Walton	Brüggen		Biber	16
			Pre-Tiglian			17
			Amstel			18

(or RECENT)

Many ocean cores have now been investigated with the help of oxygen isotopes and fossils; a picture of climate during the last million years is fairly well established, and generally consistent with the ice-ages that have been deduced from the evidence collected by geological examination on land. We have seen how, during the past 125,000 years, a fairly rapid descent into the last ice-age was followed by a correspondingly fast change back to conditions as they are today, and also that the ice-age period was interspersed with warm episodes in much the same way that our present century has evidenced runs of hot and cold summers and winters. As might be expected, the geological record varies in detail from one area to another. For example, in North America four main ice-ages are accepted, while in Britain and northern Europe seven such glaciations are recognized, and in two instances these seven are at different times. The oxygen-isotope record analysed by Emiliani provides nine ice-ages and nine interglacial episodes during the same geological period, as shown in the table above.

Opposite: freakish in its appearance and movement, a tornado also seems capricious in its effect, removing the roof and walls of a house, but leaving the remains of breakfast on a table

The even numbers in the Emiliani scale correspond to glaciations and if the reported cold periods from all areas are included, each even number in the table fits with a geologically reported ice-age at one or the other of the regions considered. The ocean–core record does then fit roughly with the evidence from land. Although the geological observations on land become progressively less certain as older events are investigated, since each succeeding glaciation messes up the remains of previous ones, the sea-bed cores are themselves not above suspicion in this respect because of disturbance by slumping and turbidity currents. However, by working on more cores and by using all possible fossils and geophysical age dating methods, and by making more measurement of sea-level changes, it is becoming possible to write the world's ice-age history back to half a million years ago or more. Careful work by such scientists as Hays, Imbrie and Shackleton (a team of American and British scientists who have studied the fine details of cores from the deep sea-bed) has provided dates for key events such as the end of the most recent ice-age and the end of the number 12 ice-age of Emiliani's classification (Fig. 19) which corresponds to the end of the Lower Pleistocene ice-age known in the USA as the Nebraskan and in the Alps as the Günz. If we take 440,000 years as the end of number 12 ice-age, we can date the beginnings of the cold periods shown in Fig. 19 as

3/2	30,000 years
5/4	70,000
7/6	194,000
9/8	294,000
11/10	364,000
13/12	476,000

provided it is assumed that the core from which the isotope ratios of Fig. 19 were obtained was deposited at an equal rate for the last half million years. This is a very uncertain assumption, but as a better understanding of sedimentation in the oceans is reached, and by comparing many different cores collected in localities where steady conditions may be expected to have existed, and by examining the cores for similarity of deposited material, it is hoped that more positive dating and more reliable correlation with the geological observations on land will be achieved. In general, as will be noted from the table comparing the geological classification with the numbered cold and warm periods derived from ocean-bed cores, the geologically observed ice-ages become subdivided when the core record is examined, and the last ice-age – discussed in detail earlier in this chapter – splits into two cold periods, starting about 70,000 and 30,000 years ago, as well as the first downward drift in temperature from the warm interglacial period, 120,000 years ago.

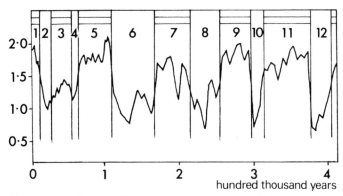

19 The ups and downs of mean temperature over the last 400,000 years, worked out from the proportions of two different oxygen isotopes in cores taken from the deep ocean. The odd numbers identify the warm intervals and the even numbers the glaciations. The vertical scale is in units which might correspond to average temperature but the exact meaning rather than the relative values are uncertain. (After Shackleton and Opdyke modified)

One of the reasons why it is interesting to probe back into these old ice-age periods is to forecast what may be in store in centuries to come. Since even in the present day, which is a warm interglacial interval, the weather in temperate zones has an enormous variation from year to year and from season to season, the world-wide ice-ages and intervening warm periods are to some extent relative terms. One thing, of course, is certain. When a vast ice-sheet spreads inexorably from the Arctic regions over North America or north-west Europe, an observable and practical change in living conditions is forced on the inhabitants of these areas. South of the ice belt there will be a large variation in annual climate, and probably also longer-term changes like the Little Ice-Age of 1400–1700. These may, if the average temperature swing is large enough and it persists for long enough, easily be reflected in changes of sea temperature and thus be recorded by the fossils found in deep-sea cores, and by the comparatively rapid response by insects on land. The existence of the large ice-sheets and of water locked up as ice in mountain glaciers on land, can be deduced from careful observations of sea-level in past geological times. When the ice-cover, the sea temperatures and the land temperatures have been sufficiently well established, it should be possible to deduce the atmospheric circulation that prevailed in past ages. Thus it may be possible to say what the living conditions were like in the land to the south of the invading ice. It is already possible to say that the onset of glaciation brought plenty of rain to the Mediterranean area, since the sea was warm enough for easy evaporation and the cyclones which bring the miserable weather to Europe were operating farther to the south. Pushing the present-day weather pattern south would mean that lakes in Africa could be maintained at much higher levels than today, and many signs point to the suggestion that, when the next ice-age comes, northern Europeans should consider emigrating to Africa.

An examination of the evidence of old shore lines, coral reefs and wave-eroded rock layers shows that the world-wide sea-level 40 million years ago was over 150 metres higher than today, and that during the severe times of the ice-ages of the last 500,000 years the level went down to 100 metres less than it is now. This means that there could not have been the two-mile-thick ice-caps that cover the Antarctic continent and Greenland, and that most of the glacier ice of the world was back in the great 300-million-cubic-mile ocean reservoir. This reading of the past is confirmed by an examination of rocks laid down around 30 million years ago in Grahamland and Patagonia, since fossil remains of leaves, wood and pollen indicate plenty of tree life in the Antarctic of those days. Cores from the sea-bed show that the East Antarctic ice-sheet started about 10 million years ago, probably helped by the greater height of the land which has since then been pushed down by the ever-thickening ice load.

The beginning of the Antarctic ice-sheet marks the start of the drift to the cold climate of temperate regions today and the intervening ice-ages of the last half-million years. It looks as if, in earlier geological times, the Earth was generally warmer than now, with corresponding changes in wind circulation and rainfall pattern. Ice-sheets were probably only present in high mountainous places, and the water removed from the sea was quite small in volume compared with what it is today.

It appears then that the massive glaciation experienced during the ice-ages of North America, Europe and Russia is geologically a comparatively recent phenomenon in the 4,500 million years' history of the Earth. This may be a useful clue when we attempt to answer the question 'why did the ice-ages happen?' Ice-ages, and the fluctuations of average temperature that occurred during them, are things that must be explained before we are capable of making meaningful statements about the world climate in the future.

6 Ice-ages: when and why?

One of the obvious places to look, if we are to explain the fluctuations of average temperature that cause the change from today's interglacial period to the last ice-age of 20,000 years ago, is the sun itself and the heat it supplies to make the Earth habitable. The sun may radiate more energy at certain times, or the heat received by the Earth may be affected by the atmosphere between the Earth and sun. Then again, as was noted in an earlier chapter, the percentage of heat taken up by the Earth's surface is affected by the albedo, or reflectivity of the different materials such as ocean, desert, forest or snow which collect the radiation. In the last hundred years, since the existence of a series of regular ice-ages in the northern temperate zones was discovered, many theories have been developed to explain the temperature changes, averaging 10° C, which have occurred fairly regularly over the past million years. Do any of these theories fit with the observed facts well enough to give such confidence in explaining the ice-age mechanism that advice can be given concerning the long-term climatic trends of the future?

The temperature at the sun's surface is about 6,000° C, while the inside, where the heat-producing change from hydrogen to helium occurs, is in the region of 20,000° C. The radiation sent to the Earth is the equivalent of 1.4 kilowatts per square metre when it reaches the Earth's atmosphere, more than enough to operate a household electric heater. Accurate measurements of the sun's heat output have been made only over the past 50 years and are not adequate to show whether there has been a significant change during the 4,500 million years' life of the Earth. There is evidence that fluctuations of a few per cent – associated with observable changes in the sun's surface appearance such as sunspots and solar flares – may take place, and these may account for climatic variations over periods of a few years. It is calculated that a decrease of the sun's radiation of 10 per cent would cause the whole Earth to be covered with ice, and if that ever occurred the Earth might never recover because most of the sun's heat would be reflected out to space instead of being absorbed to warm the Earth's surface. It is probable that the average output of energy from the sun is constant over the hundreds of thousands of years that span the ice-ages, and therefore we must look to other changes to account for the observed ebb and flow of ice-sheets in the northern hemisphere during the last million years.

Although the sun may be a regular emitter, it is possible for something to happen to the energy while in transit. This is especially so, as we shall see in the next chapter, when the sun's radiation meets the upper atmosphere. It is here that the interaction of high-energy cosmic rays and ultra-violet radiation with the gaseous atmospheric particles may alter the amount of ozone, which in turn may affect the climate at the Earth's surface. The rain of particles on the Earth is the cause of the aurorae in polar latitudes, and those beautiful and awe-inspiring atmospheric scenic effects are accompanied by rapid fluctuations of the Earth's magnetic field. Since this magnetic field has an effect on incoming particles from the sun or even from farther afield in outer space, it is possible that magnetic field reversals, which are so useful in helping to correlate the geological evidence in core samples from the ocean bed, may have altered the radiation reaching the Earth in the past. It is unlikely, however, that the type of radiation affected would have had any noticeable effect on the total heat received by the Earth, and it more probably caused biological troubles by knocking out delicate species of plant or animal and causing mutations by bombardment with high-energy particles. This could well upset some of the deductions that are drawn from the examination of core-samples, because different lines of evidence are not independent and therefore do not carry so much weight as might at first sight be supposed. One of the irritations caused by changes in the high-energy particles reaching the Earth is the variation in the amount of carbon-14 produced in the atmosphere, because this means that the C-14 time-scale is not constant, and is varying in unison with some other phenomenon whose time-scale needs to be measured.

It has been suggested that clouds of interstellar dust may have changed the sun's radiation in the past and that, closer to the Earth, abnormal volcanic activity could have cooled the Earth's surface enough to cause ice-ages. Although all the changes in the atmosphere are potential culprits, and might be blamed for the temporary glacial misery of the ice-ages of northern Europe, North America and Russia, the atmosphere is much more likely to be responsible for the shorter-term variations of climate, the occasional 'seven lean years' which are recorded in history, and are to some extent supported by measurements over the last 300 years.

There is one way in which the sun's heat could change regularly over time-spans of the order of 10,000 years or more, and it is something that is familiar in everyday experience and in the study of present-day climate described in Chapter 2. The tilt of the Earth's axis produces the change from summer to winter as the Earth travels its annual orbit round the sun. At the present time the axis is at an angle of $66\frac{1}{2}°$ to the plane in which the Earth orbits and therefore the sun is overhead at midday at the equator in March and September, at $23\frac{1}{2}°$ north latitude at midsummer in the northern hemisphere and at $23\frac{1}{2}°$ S in December. The sun in fact appears to move from the tropic of Capricorn ($23\frac{1}{2}°$ S) through the equator to the tropic of Cancer ($23\frac{1}{2}°$ N) and back as the Earth completes its annual circuit round the sun. The tilt of the Earth's axis is not constant, and is changing by about one ten-thousandth of a degree each year. Ten thousand years ago the tilt was at its maximum of about $24\frac{1}{2}°$ and is now gradually working towards the minimum figure of $21.8°$. The change in the angle of tilt of the Earth's axis goes on regularly, oscillating between $21.8°$ and $24.4°$ every 40,000 years. Increasing the tilt does not have much effect in the equatorial regions, but it means that there is a greater supply of heat in high latitudes because the poles are leaning more towards the sun during their respective summers. There will be, of course, a corresponding decrease in heat in the polar areas in winter, but since there is not much winter heat in any case, the increase in tilt of the Earth's axis favours warmer climate in the northern hemisphere land areas prone to ice-ages. The effect is not so noticeable in the southern hemisphere since the movement of ocean water tends to smooth out changes in heat supply. When the tilt of the Earth's axis is at its minimum, the high latitudes experience cold summers with average temperatures not much different from the winter temperatures. This is just the climate to favour big glacier development, with plenty of winter snow, and not enough summer warmth to melt the previous winter's deposit.

There are two other variations in the Earth's orbit that may affect the amount of heat received from the sun. The Earth's path about the sun is not an exact circle, but an ellipse, with the sun placed to the side of the mid-point of the ellipse. This means that each year the Earth approaches to a nearest point to the sun and a most distant point, and the sun's radiation correspondingly waxes and wanes. Every 96,000 years the Earth's orbit becomes very nearly a true circle, when the distance from the sun is the same all the year round. In the times of greatest deviation from the circular path, the intensity of the sun's radiation shows a seasonal change of about 30 per cent; at the present time the difference between nearest distance (called perihelion) and farthest distance (aphelion) is about 7 per cent. It will obviously make a difference to the Earth's climate if there is a large change in the heat budget during the year on account of the eccentricity of the orbit. Furthermore, the effect will depend on whether the maximum heat, which is at the nearest approach to the sun, happens to be in winter or summer. This is where note has to be taken of the third variable associated with the Earth's orbit round the sun. At the present time the Earth is nearest the sun in January, but this is not always so, because the elliptical orbit rotates around the sun and the Earth's axis, which is tilted with respect to the plane of orbit, also rotates but in the opposite direction, giving a resulting change in where summer and

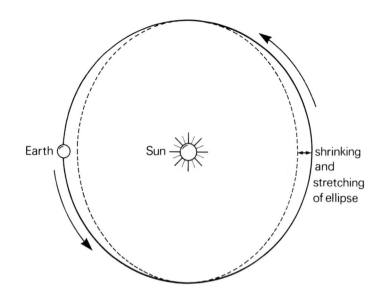

Earth ○   Sun ☀   ← → shrinking and stretching of ellipse

'wobble' of Earth's axis

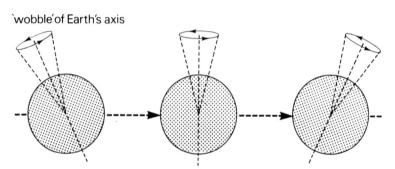

change in angle of Earth's axis relative to plane of orbit

winter occur along the elliptical orbit. In other words, every 21,000 years the nearest and farthest points take their turns regularly in the seasons of the year – a phenomenon known as the precession of the equinoxes.

    Milankovitch, a Russian expert of the traditional painstaking style, first put together the three effects – the tilt of the Earth's axis, the changing ellipticity and the precession of the equinoxes – and calculated from the known astronomical variables what would be the available heat from the sun during the past 600,000 years. This work has been extended back for one million years, and the calculations have also been made for a million years ahead. Milankovitch made the assumption that the important factor that determined ice-ages was the summer heat supply at latitudes of 60°–70° in the northern hemisphere, and on this basis his values of heat avail-

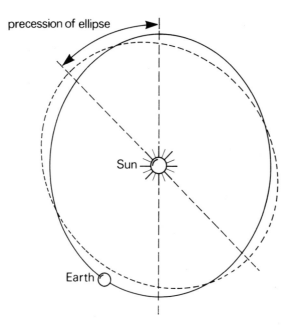

precession of ellipse

Sun

Earth

20 *Opposite and above*: The Earth's orbit round the sun shows slight
but significant changes, over thousands of years, in four respects: the
'wobble' of the Earth's axis, the angle between this axis and the
orbital plane, the change in the orbit's shape from nearly circular to
more elliptical, and the slow rotation of the ellipse itself round the
sun. Each of these affects the amount of solar radiation received, and
one or more of them acting together could have a considerable effect
– perhaps even an ice-age

ability, as calculated from the Earth's orbit changes, provided a
rough explanation of the ice-ages and the warm interglacial periods
of the past million years. There is a time-lag between the time of
maximum heat from the sun and the temperature changes that
occurred on Earth as determined from biological and other evidence (Chapter 5), because it takes several thousand years to diminish the ice-age cover of snow and ice. While the snow cover is
present, the actual heat received by the Earth is much less than that
offered by the sun on account of the high reflection of energy. If
this change in the Earth's albedo is taken into account, it appears
that the alternation of ice-ages and warm present-day climates corresponds to a world average temperature variation of about 5°C
associated with the changes in the sun's radiation caused by the odd
geometrical intricacies of the Earth's orbit round the sun.

Although it appears possible that the fairly regular series of ice-
ages which have occurred during the last million years are due to
changes in the heat that arrives at the Earth, many ingenious
explanations have been put forward based on what occurs on the
Earth itself. The growth and decay of ice-sheets themselves, since

they affect the reflected heat, may cause a cyclic freezing and melting of water in the northern hemisphere. The presence or absence of an Arctic ice-sheet and the expansion of the thick ice covering of Antarctica have both formed the central theme of geophysically orientated theories. Oceanographers have connected ocean circulation and heat absorption with climatic changes, and also with a possible effect associated with the distribution of carbon dioxide between atmosphere and ocean. Some mathematicians have even proposed that there is no point in looking for specific mechanisms for such climatic phenomena as ice-ages, since they will occur in the fullness of time independently of the initial conditions. This is the argument of the school of thought that despairs of long-range weather forecasting on the grounds that even the weather itself does not know what it is going to do next. However, with the improvement of modern geophysical techniques of dating and by using all the fossilized biological and geological evidence, it should be possible to diagnose some specific cause-and-effect relations with enough certainty to enable some tentative looks into the future.

There is no doubt that there is built-in stability in many processes that take place in the world. We have already seen that the sea absorbs the heat from the sun and does not exhibit much of a rise in temperature because the surface water mixes with the bulk of the oceans. In a similar way, the ocean water acts as a buffer to the increasing carbon dioxide in the atmosphere, by locking up the CO_2 while the ocean currents circulate slowly around the world. On the other hand, there do appear to be some signs of instability, as for example in the formation of ice-sheets which provide an increase in the reflecting power of the Earth's surface and so cause a decrease in the heat absorbed from the sun, and consequently a possible fall in temperature to encourage yet more snowfall and ice formation. As we shall see later, the growth of ice-sheets does not progress inexorably in this fashion because interesting fail-safe mechanisms come into play, and produce plausible explanations for the oscillating nature of the climate of the ice-age period. One theory of ice-ages which follows the natural stability concept assumes that greater wind circulation and hence more evaporation would follow a temporary rise in heat from the sun. There would follow an increase in cloudiness and precipitation of snow in high latitudes and subsequent formation of ice-sheets, but further increase of the sun's radiation would turn the snow to rain and melt the ice. It is interesting that warming up of part of the Earth's surface should produce glacial conditions elsewhere because of the increase of cloud, but the theory does not accord with the fact that an overall world temperature fall rather than a regional effect is the mark of an ice-age.

The continents, which cover only one-third of the Earth's surface, have changed their positions remarkably during the life of the Earth. The present Sahara desert was sited at the south pole some 450 million years ago, and rocks of this age show the typical characteristics of the recent northern hemisphere ice-ages. There are many places in the world where similar traces of glaciation exist, going right back into Pre-Cambrian times of over 2,000 million years ago. The evidence is naturally not so clear-cut as is that associated with the ice-ages of the last million years, but there are parts of present-day land which have been heavily glaciated in the distant geological past. Continental drift, together with change in position of the poles which have been tracked by magnetic measurements on old rocks, probably accounts for these old glacial periods, especially when it is allowed that the land surface can be raised during the sideways movement of the continents or drifting 'plates' of surface Earth material. When any segment of continent drifts over a polar region, or when a large area gets pushed upwards in a conglomeration of mountain ranges, then heavy snowfall and subsequent compaction of the snow to form glaciers could take place. While the modern geological theories provide reasons for the glaciations of hundreds of millions of years ago, they do not explain why, during the last million years, a series of cycles of ice-age and temperate climate have occurred in the northern hemisphere.

The peculiar arrangement of the land in the world today may be seen quite clearly by looking at a globe. Most of the land is situated in the northern hemisphere, and has almost a mirror reflection in the form of sea in the southern half of the Earth. The effect of the sun's heat, or the lack of it, is much more noticeable on land than over the water, since the land can experience large extremes of temperature, while the sea stirs its heat supply into the depths and behaves as a heat reservoir. Differences in climate in the two hemispheres are to be expected with today's land configuration, especially in the higher latitudes where there is much land in the northern hemisphere and almost free ocean in the south. At the poles themselves the mirroring of land and sea is reversed, the Arctic Ocean being reflected by the continent of Antarctica. The symmetrical opposition of land and sea may be considered too accurate to be fortuitous, but that is a geological problem. Our present concern is whether this land arrangement on the Earth has any bearing on the succession of ice-ages that have occurred in the recent days of the Earth's geological history. A shadow of what was to come was cast by the cooler days of 50–70 million years ago, when many ancient monsters which had flourished in the previous 200 million years were killed off. It was at this time, during the Tertiary geological period, from about 70 million to 1 million years ago, that the present continental arrangement settled down, with the North Pole centred in an almost land-locked sea and the South Pole in just the opposite situation, sitting in the middle of a large land mass completely surrounded by deep ocean. These peculiar conditions

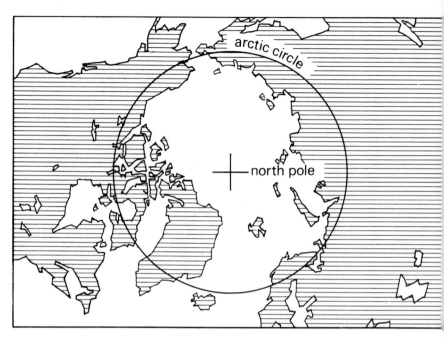

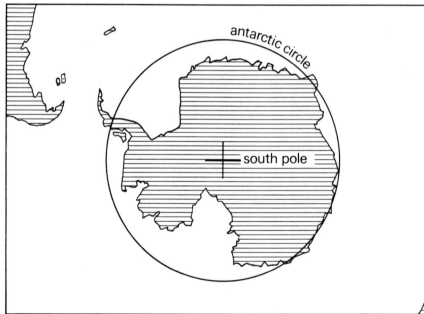

21 There is a strange and so far unexplained similarity, or balance, between the mass of land at the Antarctic and the expanse of ocean at the Arctic

in the Arctic and Antarctic have inspired scientists to formulate theories of climatic behaviour which could account for the periodic nature of the ice-ages of the last million years.

An open sheet of water at the cold North Pole would encourage the formation of ice on the surrounding land. Locking the water up as ice on land would lower the sea-level and thus block up the few passages where warm Atlantic water circulates to mingle with that of the Arctic Ocean. The Arctic Ocean would freeze over, thus removing the source of moisture which was replenishing the glaciers. The glaciers would melt and the warm interglacial climate of the northern continents would gradually ensure that the Arctic again became free of ice – and then, according to Ewing and Donn, those great geophysicists from Lamont Geological Observatory, USA, who proposed the theory, the cycle would start up again. The flaw in this theory is that the cold Arctic water, even if unfrozen, could not evaporate fast enough to supply the vast mass of snow to build up the ice-age glaciers all the way down to France and the Great Lakes. The moisture-laden air must come from the larger, warmer oceans to the south. However, the general concept of an ice-free Arctic Ocean triggering off a new regime may be acceptable, and a study of the air circulation and temperatures of the whole northern hemisphere suggests that, once unfrozen, the Arctic Ocean would not freeze over again until the surrounding continents were once more wrapped in ice. The evidence on the ground suggests that the ice cover in Alaska was formed from the south, leaving northern Alaska unglaciated, but some contribution to starting up the ice-age could have come from Greenland and northern Europe and Asia.

One interesting facet of the Ewing theory is the way it accounts for migration from Asia to South America. During the ice-age when there was an ice connection, or even a land bridge on account of the lowering of the sea-level, across the Bering Strait, Eskimos from Russia could cross to Alaska and live on the shores of the unfrozen Arctic Ocean, leaving their flint instruments behind to mark their sojourn. When the Arctic Ocean started to freeze, so the ice-sheet of North America began to recede, especially to the east of the Rocky Mountains, leaving a north-to-south migration route which was followed by the Indians right through to South America.

Antarctica today is covered with an ice-sheet which is over 6,000 feet (2,000 m.) thick and accounts for 90 per cent of the world's ice. There is evidence, from traces of glaciers on mountains that rise above the present ice-sheet, that the ice has been even thicker in the past, and that in some areas the glaciers have spilled over mountain passes to surge out to sea as a spreading ice-sheet. Robin, who pioneered much of the early measurements of both Arctic and Antarctic ice thicknesses, proposed many years ago that major ice surges might occasionally remove half of the Antarctic ice to form a vast ice cover as far north as 50° south latitude. The ice would be lubricated by water produced by the ice melting under the pressure of its own weight, the melting being enhanced by the badly conducting ice blanketing the heat flow from inside the Earth. This forms the starting point of another theory to account for periodic ice-ages. The sliding ice would be helped along its course on land by water produced by the heat of friction, and would spread out to form a large area of about 80 per cent reflectivity to the sun's heat in place of the fairly efficiently absorbing sea. There would be a 4 per cent decrease of heat intake by the Earth, which would cool the Earth as a whole, upset the wind patterns in the northern hemisphere and cause the start of a new ice-age. Eventually, since the large expanse of southern ice protects the glaciers of Antarctica from their normal replenishment of snow, the great ice-sheet would melt and allow the sun's heat to warm the sea once again.

It is noticeable that the Antarctic ice-cap does increase in thickness at times like the present interglacial period, especially in the first part of this century during a few decades of global warmth and strong wind circulation, and it is demonstrably possible for glaciers to surge out to sea at intervals of time. The criticism of the theory is based on evidence from ships' reports of ice-floes in the southern ocean at various times during the nineteenth century. There seem to have been a series of smallish surges from individual valleys rather than a massive single flow. This is supported by examination of deep-sea cores from the southern ocean which are not indicative of an ice-shelf of large extent. On the other hand, the topography of Antarctica is such that an ice movement from Wilkes Land or the Filchner Ice Shelf in the Weddell Sea could take with it about a quarter of the total ice, an amount which has taken 70,000 years to accumulate, and this would be enough to give a cover 200 metres thick up to 55° south latitude. This large flop of ice into the sea should raise the world sea-level suddenly by about 20 metres and if this was the sign of the end of the interglacial period and the birth of a new ice-age, this event would be marked by a significant advance of the sea on to the land. Such invasions are found at the end of the two most recent interglacial periods in southern England, and they are accompanied by the circumstantial evidence of remains of whole animals, perhaps as the result of a sudden catastrophic occurrence.

There is one aspect of the land and the sea which should be considered when theorizing about the formation and change of large ice-sheets on the sea. All sea water contains salt, and the salt content, together with the temperature changes, determines the water density, which is one of the controlling factors of the circulating currents of the oceans. The addition of salt to water also lowers the freezing point to 2° C less than that of water. Now the Atlantic water is more salty than that of the Pacific, and this partly accounts for the sea in the Bering Strait freezing down to latitudes of 60° N while to the north of Scandinavia the ice in winter seldom comes south of 75° N. The reason for the Atlantic being saltier than the Pacific is a climatic one. The steady trade winds carry fresh water from the Atlantic across the Panama isthmus all the year round and so cause a slight concentration of salt in one ocean and a dilution in the other. Any possible return flow of moisture by the westerly winds of higher latitudes is blocked by the Rocky Mountains and the broad North American continent. Other changes in the salt content of the sea affecting ice formation are associated with fresh water brought by rivers to the sea; these help to emphasize the complicated interaction of atmosphere and ocean which determines climate. If, for example, the trade wind decreased in the region of Central America, there would eventually be a movement of ice further south towards Norway.

While the theories of ice-ages that are concerned with terrestrial causes provide useful ways to explain automatic stability – demonstrated by the loss of moisture to supply glaciers when the surrounding ice-sheet pushes the open ocean too far away – or amplification, as when the growth of an ice-sheet is increased because of diminution of the sun's heat by the low receptivity of the ice, and although combinations of start-up and stop of ice-ages have been suggested, none of these hypotheses gives a time-scale against which the periodicity of the ice-ages can be checked. A different approach has been adopted by Hays, Imbrie and Shackleton (see p. 82) and it may prove a useful complement to the terrestrial investigations. In this work the best available modern evidence for the onset and finish of ice-ages has been carefully dated, and an analysis has been made to find the predominant periodicities in the climatic changes that have occurred over the last half-million years. The analysis was intended to show that the ice-ages were the result of variations in Earth's orbit, and it was successful in this, but, as will be seen, the most important of the variables associated with the Earth's orbit was not the one that might generally have been expected.

The experimental material consisted of two cores from the deep sea-bed, and the methods of analysis and dating were on the lines of those explained in Chapter 5. The location of the cores was almost equidistant from Africa, Australia and Antarctica so that there would be no abnormal variations due to sediment from continents. The rate of accumulation of sediment was greater than 3 cm in a thousand years so that thin enough sections could be taken to discriminate climatic changes of thousands of years. The cores between them covered a time span of 450,000 years.

The reason that the cores selected were from the southern part of the Indian Ocean was in order to follow changes in the sea temperature world-wide rather than to measure temperature fluctuations near to the ice-age happenings of the higher northern latitudes. Three different measurements were made throughout the cores. The oxygen-isotope ratio (see p. 61) provided a measure of the amount of the ocean water locked up in the northern hemisphere ice-sheets, since the Antarctic ice was more or less constant during the period under investigation. The summer sea-surface temperature at the core sites was obtained from examining the assemblages of microscopic skeletons of animals called radiolaria. The third investigation was of a particular type of radiolarian which thrives in special conditions of salinity and temperature in the surface waters and is especially abundant in glacial times.

The oxygen-isotope measurements showed a pattern similar to that for the Pacific cores (p. 82), going back two million years. The results of the Pacific investigations were used to provide an accurate date of 440,000 years ago for a clear change in oxygen-isotope ratio, and the date for the Pacific core was based on age-dating by potassium-argon measurements of a conspicuous reversal of the Earth's magnetic field. Other date checks were made at well-marked places in the cores, which were established by much previous core analysis at 117,000 and 251,000 years, and the time-scale of the whole length of core was fixed by assuming a constant rate of sedimentation in between the check points.

The three indicators – oxygen isotope, surface water temperature and the special radiolarian percentage – gave similar, but by no means identical pictures when plotted against age. A frequency analysis was made of all three sets of results, and in all cases three main time periods were apparent, at around 100,000 years, 42,000 years and 22,000 years. Earlier in this chapter, it was noted that the eccentricity of the Earth's orbit changed from almost circular to maximum with a cycle of 96,000 years, while the tilt of the Earth's axis oscillated a few degrees every 40,000 years and the precession of the seasons around the orbit took about 21,000 years to complete a circuit. Although it seemed probable that the tilt of the Earth's axis would be the most important Earth orbit phenomenon determining ice-ages, because the snowy winters and cold summers in northern latitudes were expected to increase the ice-caps by feeding the glaciers and not melting them in the summer, it turned out that the 100,000-year time period was the predominant one, indicating that the change in the Earth's distance from the sun was more important than the angle at which the polar regions faced the sun.

There is no doubt in the minds of the authors of this study of variations of the Earth's orbit that, by their careful selection of sea-bed cores and by their painstaking measurement of minute fossils and oxygen isotopes, they have discovered meaningful results. In the summary of their report, 'It is concluded that changes in the

Earth's orbital geometry are the fundamental cause of the succession of Quaternary ice-ages.' The measurements extended over about 450,000 years, about half the recent ice-age time-span, and therefore only included a few of the largest period-dominant time cycles related to the eccentricity of the Earth's orbit round the sun. It is possible that, although the change in distance between Earth and sun is the major factor that ensured an ice-age during the last million years, other factors played their part in determining the severity of any particular occurrence. For example, it perhaps needed an almost land-locked Arctic Ocean to allow an ice-age at all; or the periodic mechanism of the Antarctic ice-cap growing and sliding off to make a greatly increased white expanse in the southern hemisphere is a necessary adjunct to the change of seasonal sun's heat due to the variations of the Earth's elliptical orbit.

There may be a tendency to amplify the effects of the variations of the heat received by the Earth on account of orbital changes because of some changes in the sun itself, or in the Earth's atmosphere. We shall see in the next chapter that changes do occur in the output from the sun, and that volcanic dust and possibly changes in cloud cover could affect the heat received at the Earth. It is even possible that a change in the sun's emission could start off changes in the Earth's atmosphere, and so accentuate, into something noticeable to the world's inhabitants, what might otherwise be only a minor change in Earth temperature. There is strong evidence that short-term changes such as the 11-year sunspot cycle affect the climate to some extent, and this is why care must be taken to understand all the other associated effects, including those produced by human activities. In the meantime, it appears that the ice-ages which were really severe for the northern temperate regions were associated with the orbit of the Earth round the sun. There is an important deduction to be made here. Since the astronomical figures are calculable, and repeat at regular intervals, there could well be more ice-ages to come.

7 Heat from the sun

The sun is a typical moderately sized star, with a diameter about 100 times that of the Earth. Although the sun is 93 million miles from the Earth, it is hot enough and large enough to be the primary source of energy for the Earth. This energy is produced by the fusion of hydrogen nuclei to form helium, at the rate of 564 million tons of hydrogen a second; this is only a very small fraction of the total, so that there is no fear that the supply will run out. In fact, it is probable that the energy from the sun is constant except for bursts of energy in the form of electrically charged particles, X-rays and ultra-violet rays at the time that dark patches or 'sunspots' are seen on the sun's disc. The effect of the high-speed particles is to form the impressive auroral displays that are a feature of the higher-latitude skies, and which are caused by the excitation of atoms of gases such as hydrogen, nitrogen and oxygen in the rarefied upper atmosphere of the earth, just as yellow street lamps are the result of electrical discharge in the presence of sodium vapour, and the blue lights are similar effects with mercury.

The extra solar energy emitted at the peak of the 11-year cycle of sunspot activity is only about one per cent above the regular output. Although there is some circumstantial evidence of a connection between sunspots and droughts in the dust-bowl zone of the USA, and also with winter rainfall in the United Kingdom (which since 1873 shows a $7\frac{1}{2}$ per cent increase above average just before the sunspot maximum and a similar decrease before the minimum) the consensus of opinion is that any slight variations there may be are not easily separated from the larger year-to-year fluctuations which arise from other causes. There have been suggestions of periodicities of 80–90 years and 180–200 years, based initially on early observations of sunspot activity, and checking vaguely with records from oxygen-isotope measurements in Greenland ice-cores, but it does not appear to be proved that change in the sun's energy output has had a systematic effect on air temperature and circulation. A slight trend towards increase of the solar radiation is suggested by direct measurements at the Smithsonian observatory between 1921 and 1952, although the average radiation was constant to within 0.2 per cent. This extra energy could be the cause of the warming-up period of the first half of the twentieth century, but until much more extensive sets of measurements are available to check with climate observations, it is best to consider the sun's energy to be constant, except for the odd bursts of activity at sunspot time and perhaps the odd 'flicker' mentioned in the next chapter to account for geological effects over a cycle of about 250 million years.

Part of the sun's energy penetrates through the Earth's atmosphere and reaches the surface in the form of sunlight, but some of it is caught by the atmosphere or trapped in the space above the atmosphere. High-speed protons and electrons tend to get trapped in the Earth's magnetic field in the region from a few hundred to a few thousand miles up, and they spiral around the Earth in the Van Allen belts which were one of the great satellite discoveries of the 1960s. Some of the particles may spill over into the upper atmosphere and here they could trigger off changes in ionization; they may be the cause, for example, of changes in the production of carbon isotopes (C-14), and they may be connected with the climate changes associated with sunspots because they could possibly cause trends towards patterns of cyclones or anticyclones. However, while the radiation belts may be blamed for some of the weather, or day-to-day changes in temperature and rainfall, it is in the atmosphere itself that the long-term interaction between the sun's radiation and the Earth takes place and where changes in climate are determined.

The Earth, in its journeying through space, carries with it an envelope of gas which is held fast by gravitational attraction. The greatest concentration of the atmosphere is near the Earth's surface and the air gradually becomes more rarefied as the height above the surface increases. At sea-level, the weight of the atmosphere above us is 14.7 lb per square inch, the familiar 'atmospheric pressure' (in the metric system 1,000 millibars), while in the ionosphere, 30 to 300 miles up (50 to 500 km), the pressure is about a hundred-thousandth of this value. The composition of the atmosphere is four-fifths nitrogen and one-fifth oxygen, with an odd one per cent made up mostly of argon, but with small traces of other rare gases, as well as carbon dioxide, methane, hydrogen, ozone and nitrous oxide. Above 70 miles (100 km) what little material there is consists mostly of oxygen, with nitrogen in the form of nitrous oxide. In this region, X-rays and ultra-violet radiation are used up in stripping electrons from neutral atoms to form layers of charged particles, which form a reflecting surface for radio waves.

Further shielding of the Earth's animal population from deadly X-rays and burning ultra-violet is provided by the stratosphere. In this layer, there is a peak concentration of ozone around the 12-mile (20–25 km) level, formed from oxygen by the active ultra-violet radiation, which is itself absorbed in the process. The ozone is broken up, as well as being produced, by the radiation, and a balance is achieved, with less ozone at higher temperatures. Some ozone drifts down to lower layers in the atmosphere, but continual replenishment takes place in the stratosphere. It is possible that the ozone behaviour provides a link between the activity of high-energy particles in and above the ionosphere and the lower atmosphere; this could account for the supposed connection between sunspots, which affect the sun's high-frequency radiation, and the weather at the Earth's surface. The stratosphere has been investigated by instruments carried in balloons and satellites and it has been found that, although the atmosphere normally gets colder as one goes higher (about 3.6° F per 1,000 feet, or 6.5° C per km) at the top of the stratosphere 25 miles (40 km) up, the temperature rises to about 0° C, because of the energy produced while the ultra-violet rays are being absorbed by oxygen and ozone. Although there is not much air around in the stratosphere, there may be linkages with the temperature and circulation changes in the troposphere, which may have a significant effect on climate. This is why it is necessary to follow what is taking place all the way up from sea-level to outer space.

The troposphere is the lowest layer of the atmosphere and the one that we know best. It is here that clouds form, winds blow and three-quarters (by weight) of all the gases of the atmosphere, together with almost all the dust and water vapour, find their home. The lower atmosphere is to some extent boxed in by the 'tropopause', the level at which the steady fall in temperature with height gives way to a smaller or even zero temperature change. This has the effect of limiting the convection currents in the air and, to a large extent, sealing off the weather zone of the atmosphere. The tropopause moves with the season of the year and with the daily changes in barometric pressure, and is twice as high at the equator (16 km) as at the poles. In the mid-latitudes, the tropopause has to straddle the two extremes and it is here that interchange can occur between stratosphere and troposphere, with traces of water vapour going upwards and dry, ozone-laden air coming down. The troposphere forms the third shielding layer from the sun's rays, since it is effective in absorbing part of the heat from the sun, apart from providing shade with its cloud cover.

The troposphere acts also as a protective blanket to the Earth by retaining at night the heat collected from the sun's rays during the day. The Earth re-emits the energy it receives from the sun, but being a comparatively cool body, it does so mainly in the infra-red region of the energy spectrum. Infra-red radiation is strongly absorbed by water vapour, carbon dioxide and ozone, and these cause the atmosphere to act in much the same way as the glass of a greenhouse. The heat from the sun, being short-wave radiation in the visible region of the spectrum, is transmitted, while the out-going radiation is absorbed, so that the heat is retained by the atmosphere, and in turn radiated back towards the ground. Any substance that absorbs infra-red will provide this blanketing effect and

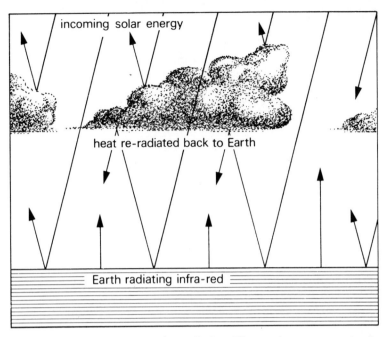

incoming solar energy

heat re-radiated back to Earth

Earth radiating infra-red

22 Incoming and outgoing solar radiation. The total amount received
from the sun is balanced by the total amount reflected back into
space from the atmosphere and clouds, absorbed by the land and sea
surface, and reflected from the surface in the infra-red waveband

will slow the cooling of the Earth's surface, while at the same time
stabilizing the atmosphere and thus reducing convection, cloud
formation and rain. This reduction of convection is what the greenhouse is best at in its work of keeping the heat where the gardener
wants it; however, the glass of the greenhouse does also play some
part in transmitting sunlight and in preventing low-temperature
radiation from escaping, and it allows us to use this simple analogy
to describe the process that takes place in the atmosphere.

Water vapour and carbon dioxide are very similar in their absorption of infra-red energy, but their overall behaviour with respect to world climate is different because of the different distribution of these two constituents of the atmosphere. Carbon dioxide
is evenly mixed throughout the air column, whereas water vapour
tends to gather in the lower parts of the atmosphere, and with its
continual evaporation, condensation and precipitation it behaves
in a much more regional manner. Carbon dioxide is absorbed by
the water of the oceans and taken up by vegetation; in general,
if there was a world-wide increase or decrease in concentration
spread over a long time, this could possibly be responsible for climatic changes. The effect of carbon dioxide absorption will be
more noticeable in cold, dry climates and in places of high elevation, since it is here that there may be a shortage of that other good

heat absorber, water vapour. A warming up of the Earth due to increase in carbon dioxide might then affect the lands in high latitudes rather than those of the tropics, by producing warmer winters in those places where some such effect might be appreciated.

There seems no doubt that the carbon dioxide content of the air has increased by 10 to 15 per cent during the past hundred years. Careful measurements, started a few decades ago, indicate that nearly half the carbon dioxide produced by the use of fossil fuels and other minor sources, such as burning lime or emissions from volcanoes, stops in the atmosphere, while 20 per cent is taken up by the oceans and 40 per cent is absorbed into plants and forests by the 'land organic pool', a large part of which consists of organic humus in the ground. There is a strong impulse at present to study the climate reaction to change of carbon dioxide, since the world is at an energy crossroads. During a period of a few hundred years, a large amount of coal and oil will have been burnt in the change-over from man- and horse-power to nuclear and solar energy supply. This fossil fuel was 'banked' by natural processes that took place over millions of years, and man is proposing to cash in this capital in a very short space of time. It would be advisable to spend a considerable effort on climate research to ensure that no ill effects are produced.

The part played by the sea in mopping up excess carbon dioxide has been measured in recent years with the help of the nuclear tests made in the early 1960s. These explosions produced quantities of tritium (a radioactive isotope of hydrogen with two neutrons in addition to its single proton) and other trace substances which probably behave in water in much the same way as does dissolved carbon dioxide. During the course of 12 years, the tritium has mixed, by a combination of convection and diffusion processes, down to a depth of 360 metres; this figure allows us to calculate with reasonable certainty how long it takes for material mixed with the sea surface to become spread evenly through the whole of the ocean waters. The mixing time is roughly 500 years, and this leads to the figure of one-fifth of the carbon dioxide being taken up by the oceans. These figures will be changed, of course, if the carbon dioxide causes a warming up of the air and hence of the sea water; a warmer sea will mean less carbon dioxide absorbed, and a further increase of air temperature. Fortunately, there are other trains of events which work towards maintaining the status quo, rather than triggering off an explosive chain reaction. It is really a full understanding of all these interesting mechanisms that concerns the meteorological modellers who are endeavouring to enlist the large capability of modern computers in order to determine what we can and cannot safely do to our world environment.

The forests take up about twice as much of the excess carbon dioxide as the oceans, and play a great role in stabilizing the atmosphere and its blanketing effect on the Earth's surface. However, the world is in an unprecedented period of population increase, and as man demands more space and more food, it may be necessary to cut down forests and replace them with houses, and with other vegetation which is not so effective in taking up carbon dioxide. It is interesting to remember that tropical forests cover an area of one-sixth of the land surface, but contain nearly half the carbon that is locked up in vegetation. An increase of average temperature might cause more rainfall and augment the growth of these forests and of other cultivation. However, any change in global average conditions may have adverse effects on agriculture, because of a shift in the location of climatic regions, and delay in people moving to the new fertile areas and adapting crops to suit a new climate pattern. These phenomena are not new in human history: we have seen in earlier chapters how human beings had to leave previously fertile sites in the Middle East, or were driven from northern temperate zones by the advance of the ice-ages.

How great is the effect of carbon dioxide in the atmosphere? This may be gauged from the estimated total 'greenhouse' blanketing of carbon dioxide and water vapour together, which is such that the average temperature of the Earth's surface is about $30°C$ higher than it would be without the atmospheric cover. If the carbon dioxide content of the air were to double there might be a $2°$ to $3°C$ rise in the Earth's average air temperature. This rise would be greater at high latitudes and might change the balance of ice in the polar regions.

If *all* the Antarctic ice and the Greenland ice-cap were to melt, the sea-level would rise about 100 metres, drowning many of the major cities of the world, and making vast swamps out of much of today's thickly populated areas. However, a four-fold or eight-fold increase of carbon dioxide would be needed to produce these catastrophic changes to the face of the Earth; the estimated temperature rise would average $6°$ to $9°C$, which is of the same order as the highest temperature rises that occurred in the hot eras of the geological past some hundreds of millions of years ago. There is no immediate call for alarm and despondency, since there will be a large increase in evaporation if the temperature of the air rises, and the possible melting of the polar ice may be counteracted by much greater snowfall, which will cause the ice to persist in spite of the higher average world temperature. It is possible that the present configuration of the land and oceans of the world is one which favours polar ice-caps; at present we recognize the existence of many conflicting forces, but are not yet knowledgeable enough to say which will control the climate, or whether there is so much in-built stability in the system that climate change is slow to occur and is marginal in its effect (except for those who are close to the margins).

An increase of moisture in the air will lead to greater cloud formation, so that automatically more of the sun's heat will be radiated. Actually, the matter is not quite so simple as this, since an increase in low cloud cover could be associated with less cloud in the upper regions. The oceans themselves provide another stabilizing influence, apart from taking up some of the excess carbon dioxide which would otherwise warm the atmosphere. The circulation of the ocean currents mixes the heat received from the atmosphere at the sea–air interface with the whole body of the ocean water and smooths out the more rapid changes of temperature that occur in the air itself. Oceanographers and meteorologists have scientific problems in common, and large-scale joint experiments are in hand to find out all about air–sea interaction, and the relative amounts of heat transferred from equator to pole by the circulation of the atmosphere and by the much slower ocean currents, in which the greater moving mass and larger heat capacity compensate for the lower speed. Further complications in the polar regions arise from the formation of glaciers and ice-caps by precipitation of snow, from the melting of the ice by the sea and from the initiation of circulation of the water by the sinking of dense cold water masses and of water with high salt content. The lack of a large enough computer to take account of all these interconnected phenomena is one of the reasons why it is not possible to predict the future climate.

At the same time that the burning of fossil fuels is contributing to a warming of the air by increasing the carbon dioxide content of the atmosphere, another infra-red absorber is being emitted: smoke and dust particles. The effect of these aerosols is both to reflect and scatter sunlight away from the Earth and to absorb energy from the sun. It is not certain to what extent man is competing with natural sources in the production of dust particles, and the effect on the temperature of the Earth is by no means so clearly established as that caused by carbon dioxide emission.

Volcanic activity, especially in the tropics, produces a world-wide increase of the aerosol content of the atmosphere. It is estimated that the eruption in 1963 of the volcano Mount Agung in Bali reduced the solar radiation that reached the Soviet Union by a significant five per cent, an amount that might have a world-wide effect at times of exceptional volcanic activity. Other natural aerosols are associated directly with the climate: sea spray and dust clouds, for example, are raised by storms and may well be extensive and persistent enough to produce regional changes, or to act in a way which reduces or accentuates the natural effects they are associated with. Generally, the aerosols are short-lived and often remain in the atmosphere for no more than about a week, so that much of the low-level dust and pollution is rapidly washed from the air before it causes much harm. The use of satellite observing stations to measure what quantity and quality of radiation is entering and leaving the atmosphere, together with regular observations of the reflected energy from snow, cloud and various land surfaces and of the growth and recession of ice-layers, should enable a distinction to be drawn between the causes of short- and long-term climatic changes. It is opportune that new methods of observing planet Earth are available just when they are needed. The fluctuations of the sun's heat output, variations of the heat that is reflected away before being used on Earth, and the greenhouse effect could all change the mean temperature of the Earth, but we do not know any of these properties very accurately. Energy is being used at an ever-increasing rate, and more of the Earth's land space is being used for man's activities so that it is necessary for the human race as a whole to find out what factors do control the climate, so that ill effects can be avoided.

One of the ways to discover more about catastrophic changes that may take place is to delve into the past history of the Earth by examining the geological history written in the layers of rock that make up the Earth's surface. These rock layers are the result of the interaction of the atmosphere with the water of the sea and the solids of the Earth, and this interaction has affected the atmosphere itself in the course of the 4,500 million years of the Earth's existence, particularly in one notable instance. Most animal and plant life today gains its energy by changing carbon to carbon dioxide with the help of the 20 per cent of oxygen in the air. Things were not always this way, however, and for the first 4,000 million years the oxygen in the atmosphere was probably less than one per cent. The atmosphere and the oceans were formed by the exhalations of volcanoes, and since the earliest days of the Earth, these have consisted of water, nitrogen, carbon dioxide, hydrogen, sulphur dioxide and small quantities of methane, ammonia and other gases. The water naturally supplied the seas and the gases formed the atmosphere. A modifying effect of the atmosphere is the loss of some of the lighter gases such as hydrogen, but the main point to note is that the volcanoes did not provide any oxygen. Some oxygen was produced, however, from the earliest times by the break-up of water vapour by ultra-violet rays, since there was no protecting blanket of ozone in the stratosphere. The proportion of oxygen was very small, because it automatically cut off its own production by absorbing the ultra-violet when a concentration of about 0.1 per cent was reached.

In order to get an atmosphere with 20 per cent of oxygen, it is necessary for plant and animal life to break up carbon dioxide in the living process of photosynthesis. Plants and animals need visible light and are not restricted by the ultra-violet absorption of ozone and oxygen which limits the amount of oxygen formed from the break-up of water molecules. It is probable that early life began in ponds where the frail growth could remain at a depth of ten metres or more, out of reach of destructive ultra-violet rays but taking in all the visible light needed for photosynthesis. Gradually the oxygen produced by the plants themselves built up a concentration in the atmosphere to provide the protecting blanket of ozone that cut the ultra-violet down so that life could carry on without the shielding water layer of the pond.

The geological history which is discussed in the next chapter shows that there were probably several steps in the upward trend of oxygen concentration. Each step provided better protection against ultra-violet rays, so that life could eventually come out of the water environment and bask happily in sunlight on dry land. It is probable that in the Carboniferous period the oxygen level overshot its present-day value. This overswing, with photosynthesis feeding on carbon dioxide and generating large quantities of oxygen, would not last too long, since as the carbon dioxide decreased, the reverse greenhouse effect would come into play, the Earth would cool and life would ease up. The overswing period of the Earth's atmospheric history did provide that large sum of energy capital in the world's bank, in the form of coal, which our present generation is cashing in at great speed. We must not forget to spend some of this capital on learning about the natural processes that maintain our balance in the world, so that we do not wreck the system.

8 Climate and the drifting continents

It is not a bad idea, when thinking of possible climatic changes in the future, to take a look into the past. We saw in the last chapter how changes in the atmosphere and in the heat output from the sun could affect the temperature of the Earth's surface. It is now time to see what has been happening on that surface in the last 4,500 million years since the Earth solidified from a molten mass. What can we learn from the study of rocks by geologists who have been reading past history from the evidence that is set out on the land all around us?

One of the fortunate aspects of geology is that, in different parts of the world, at different heights above sea-level and in various geographical configurations subjected to a whole range of climates, almost everything that could happen to rocks is to be seen in process today. Rivers carry the debris washed and ground off hills and mountains by rain and glaciers, and deposit the material in shallow seas to form fresh layers of sedimentary rock. Animal life in the sea dies and its skeletons and shells fall to the sea-floor to add to the sediment layers. In other parts of the world, volcanoes pour out molten rock which solidifies to provide layers of basalt and other igneous rock material. In general, the more recent layers of rock cover the older ones so that by observing where the rock strata have been laid bare by erosion, as in river gorges such as the magnificent Grand Canyon, it is possible to turn back the pages of geological history, layer by layer, like counting tree-rings to probe back into the prehistorical past. In much the same way as the carbon-14 isotope has been used to date the tree-rings, so other laboratory measurements may be made on suitable rock layers to determine their ages and so provide a time-scale in millions rather than hundreds of years.

The age when the oldest rocks solidified is determined by the decay of radioactive uranium, which disintegrates spontaneously to yield eventually lead and helium, both of which latter elements may be used to give a figure for the age of the Earth. Other radio-active materials useful in making a time-scale for rock strata are potassium, which has a rare isotope, potassium-40, which decays to the inert gas argon, and rubidium which changes to strontium. Both these decay mechanisms have been used extensively to find the ages when rocks crystallized, and provide a useful numerical element to the tabulations of geological strata which have been arranged in order of age, layer upon layer, in the Earth.

A careful examination of the texture and composition of the rocks themselves, together with observations on rocks being formed today, provides us with a tool for finding what was the typical climate in the past when the rock layers were deposited. The recent extensive drilling of the North Sea to discover oil and gas provides many examples of the relation between the climatic conditions that existed when various strata were deposited and the rock material itself. The gas in the southern part of the North Sea was formed during the Carboniferous period, when hot, swampy, freshwater inland seas spread from present-day England to Germany. Primitive trees and ferns were abundant and in the course of millions of years, after burial by subsequent layers of sediment, these vegetable remains formed seams of coal. Methane, produced by decaying vegetation, escaped and was held in a porous sandstone formed in desert conditions, where sand dunes like those in the Sahara today were blown and drifted before they eventually consolidated to form a thousand-foot-thick layer of Permian sandstone. Changes in the inside of the Earth then caused the whole area to sink slowly, and salty sea water broke in from the ocean in the north. The hot climate caused rapid evaporation, such as is taking place today off the west coast of India, and thousands of feet of rock-salt layers were deposited. This impervious material satisfactorily trapped the gas in the porous Permian sandstone layer so that it was preserved for our use more than 200 million years later. The shallow sea was finally replaced by a sea which was similar to, but more extensive than, the North Sea of today. From 180 million years ago, when continental movements were producing the Atlantic Ocean by the drift apart of America and Europe, the associated strains caused the North Sea area to continue sinking and to form a convenient receptacle for the sediment that was being weathered from the mountains of north-west Europe. The deposits between Norway and Scotland were mixed with the remains of marine life to form the rich oil deposits of the North Sea which have remained undiscovered until recently.

The cold and inclement conditions in which the oil companies have to work are a complete contrast to the pleasant tropical climate that existed when the oil was formed. Part of the deterioration is probably due to a general world cooling, but a large part must be due to the moving of the continental masses and their associated shallow continental shelves to northern latitudes during the past hundred million years. The movements of land about the surface of the Earth are probably caused by temperature effects below the crustal layer, which is only 20 miles or so thick beneath the continents. The depressions of a few miles which are filled with sediment, or the mountain ranges which at most are only a few miles high, are really only wrinkles on the surface of the Earth, which is a sphere of 8,000 miles diameter, and it is not surprising that they occur when forces are acting which can drive vast areas of continent apart. The effect of the internal forces on the geological picture is supplemented by the action of the atmosphere, which plays its part in the denudation of the Earth's surface features and in controlling the temperature and rainfall conditions which, together with the height of the land, determine the type of animal and plant life that can exist.

The examination of fossil remains has provided a vast amount of information about the distribution of life in the past, and thus about the climate in times going back to about six hundred million years ago. It is convenient to refer to the geological names of the periods of rock formation, but for those who prefer dates, tentative ages have been added in the following list:*

Pleistocene	Ice-ages in northern latitudes	1–3 million years ago
Pliocene	Cool	3–10 million years ago
Miocene	Moderate	10–25 million years ago
Oligocene	Moderate to warm	25–40 million years ago
Eocene	Moderate, becoming warm	40–70 million years ago
Cretaceous	Moderate	70–130 million years ago
Jurassic	Warm and equable	130–180 million years ago
Triassic	Warm and equable	180–230 million years ago
Permian	Glacial, becoming moderate	230–270 million years ago
Carboniferous	Warm, becoming glacial	270–350 million years ago
Devonian	Moderate, becoming warm	350–400 million years ago
Silurian	Warm	400–440 million years ago
Ordovician	Moderate to warm	440–500 million years ago
Cambrian	Cold, becoming warm	500–600 million years ago

The present-day climate is only the second example of glacial conditions during the last 600 million years, and the general picture is one of warm or moderate conditions. For those who feel hard done by when they see this list, it must be remembered that the earlier animals that inhabited the earth were fishes and, later on, reptiles, both of which could survive what might have been at times a rather too hot climate. The fish of Devonian times had the sea to keep them cool, and the reptiles were well designed to cope with the drying up of the ponds in which they lived, by hibernating in the mud. Perhaps mankind has, on the average, got about the right variability of climate for his well-being; on the other hand, it may be more correct to say that he has adapted to what has been handed out. For example, while the amphibians such as the frog

* Adapted from Lamb, *Climate: Present, Past and Future*, vol. 2, 1973, and Brooks, *Climate Through the Ages*, 1949.

family had to return to the sea to lay their eggs in a similar way to the fish, the reptiles learned to lay eggs with protective shells so that their young could be hatched out in dry conditions. The turtle today lays its eggs on the beach even though it spends most of its time in the sea.

The lung fish of today does not resemble exactly the air-breathing fishes of Devonian times, 400 million years ago, but it has strong survival value in that it can put itself into what is almost a death-trance in order to keep going in adverse climatic conditions (a capability to be envied by those who consider freezing their bodies in order to return to live in future centuries). With all these safety devices which the ancient animals developed, their presence as skeletons in various geological rock formations provides, in general, useful knowledge of the climatic conditions that existed when they lived. The early amphibians must have needed a reasonably warm climate since they are so similar to their successors of today, and the fact that their remains are found in Devonian sediments in eastern Greenland suggests that the climate at that time was much warmer than it is now. This warm period appears to have continued through the hot, swampy conditions during which the coal of the Carboniferous period was laid down, and the early land-living animals, which in turn led to the development of the reptiles, enjoyed the mosses and ferns of the lush, steamy subtropical forests.

There is evidence of extensive glaciation in what is now the southern hemisphere as the end of the Carboniferous period approaches, some 270 million years ago, but the late Carboniferous plants and animals of the northern hemisphere still indicate that swamps and warm environments were the prevailing regime for northern Europe and America. The beginning of Permian times was accompanied by mountain-building and a general disturbance of the land portion of the Earth's surface. Some of the glaciation whose geological signs are preserved in the rock formations may have been caused by mountain snow and ice. Some of the apparently discrepant observations on remains of life may have been the result of sideways movement of the continents rather than a real ice-age such as has been experienced all over the world during the last million years. Whatever the cause, the Earth, at the close of the 100 million years or more of warm, equable climate, was becoming a planet with much greater variability in temperature and rainfall, and in the shape of its surface. This change from a 'one-type' environment to one in which many different kinds of weather and terrain were available was probably the cause of the upsurge of the reptiles and also of their interesting diversity of size, shape and living habits, which were, no doubt, evolved in each case to meet a particular set of living conditions.

We have seen how the drilling for North Sea oil demonstrated the extensive spread of the thick sandstones of the early Permian

(270 million years ago) over north-west Europe. Evidence of hot desert conditions, both in Permian and in the following 50 million years of Triassic times, is found in deposits of red rock as far afield as Texas, Africa, China, northern Europe, India and Brazil. This type of rock is probably formed as a result of alternating periods of heavy rain and desert-like drought, and there is little doubt that, as in the North Sea area, the Permian time was generally one of hot and often dry weather, with the reptiles and their ancestors, the amphibians, revelling in the warm climate. The known behaviour of this group of animals today indicates that the environmental conditions of 250 million years ago were not unlike the delta areas of Louisiana and Texas. This is not surprising, since if the geological picture of mountain-building at the close of the Carboniferous period is correct, it would be followed by rapid erosion and the deposition of the spoil around the mouths of the rivers which carried the debris from the eroding areas down to the sea.

There was a different distribution of animals in the Karoo of South Africa and in the Dvina river area of northern Russia. The land was probably elevated several thousand feet above sea-level and encouraged the type of predator and prey animals that are found in the subtropical parts of Africa today. The change in elevation rather than large difference of temperature from equator to pole probably explains the climatic information which can be deduced from the fossil remains of Permian reptiles. This smaller temperature difference between the polar regions and the tropics may have been due to the lack of polar ice such as today covers the Arctic and Antarctic regions. Without the large ice-sheets, the behaviour of the prevailing wind pattern, initiated by the rotation of the Earth, would be different from that which we experience now, and there would also be changes in the pattern of surface currents such as the Gulf Stream, which in their time affect the more local form of the winds in the shape of cyclone and anticyclone disturbances.

The small animal and plant life which is found in the oceans, and can be traced back through geological history by microscopic examination of cores of sediment taken from the deep-sea floor, shows an almost catastrophic change in the closing years of the Permian period. Three-quarters of the minute foraminifera and other beings whose shelly fragments help to tell the climatic story of the past disappeared, together with many molluscs and types of coral. This is the most spectacular example of the extinction of fossil species, even if it did occur among the more minute forms, and it has been called a 'crisis in the history of life'. The good side to the story is that those 25 per cent of species that did survive seem to have thrived in the following Triassic period, as so often is the case when some catastrophe removes the opposition and leaves a wide-open ecological space for expanded growth.

The large reptiles of the Permian show that warm conditions existed over much of the Earth. The same can be said of the following 50 million years, the Triassic period, when the climate, judging by the animal remains and the geological story, was one of warmth with alternations of wet and dry. The moisture is shown by the widespread existence of specialized amphibians, which need streams or ponds near by, together with a generally moist climate. The mild climate, and the abundant food supply which must have accompanied it, were just what was required by the dinosaurs, which made their debut on the geological stage at this period with specimens up to 20 feet in length.

The dinosaurs were the 'typical' animal of the Jurassic period (130–180 million years ago), and many varied types of these giant animals existed for 150 million years before meeting some sort of 'Waterloo' at the close of the Cretaceous period. The heavily armoured *Stegosaurus*, with the large triangular spikes on its back, ate mostly grass. The best-known dinosaur, the 85-foot *Diplodocus* with its long neck, 45-foot tail and small head, also was a vegetarian, and like others of the species, supplemented a small brain with subsidiary concentrations of nerve centres in his shoulders and backside. Some of the dinosaurs, like *Tyrannosaurus*, stood up like a kangaroo to a height of 20 feet; with its massive jaws and sharp teeth it was probably the largest flesh-eater ever to roam the earth. *Tyrannosaurus* was probably too greedy and ate up the opposition; from the fossil remains, he appears to come towards the end of the line of these enormous animals.

It is probable that the Jurassic period was one of topographic and climatic smoothness, with low lands resulting from the previous mountain erosion, and shallow seas advancing over flat continental platforms. The luxuriant vegetation, and the natural propensity of animals to operate on the 'big fish eats little fish' principle, ensured the survival of the large and greedy, so that giantism was the order of the day, even in the case of the earlier animals such as the pterodactyls, which took to the air like great bats with 25-foot wingspans. There was probably not much differentiation between the seasons during the Jurassic period, and the average climate was tropical or subtropical, rather like the coastal areas of East Africa. Crocodiles and turtles thrived in this era, and the first type of bird, the *Archaeopterix*, with teeth in its beak, feathers on its wings and a lizard-like tail, heralded a new class of animal. The warm seas encouraged ichthyosaurs and plesiosaurs, marine reptiles whose remains have been found in the Jurassic rocks of eastern Greenland, at present in 70° north latitude.

All good things come to an end, even with dinosaurs, and the Cretaceous period ushered in a time of geological change to replace the benign, settled conditions of the Jurassic. The grass-eating giants became smaller, probably because the plant life was not growing so well, but possibly also because the carnivorous reptiles of the *Tyrannosaurus* kind were coming into their prime. Some of the evidence from the remains of plants suggests that clearly marked alternating seasons, such as are familiar in the world today, were taking the place of the more uniform tropical regime. The giant reptiles may even have died out because they were accustomed to breathing air with an excessive amount of oxygen, which was the result of the upsurge of plant life in the Carboniferous and following periods. The animal fossil remains show a tendency towards an essentially present-day spectrum of land-living animals such as snakes, lizards, turtles and crocodiles, while the now extinct dinosaurs and pterosaurs were waning both in size and in numbers. Cold climates were, however, still not widespread, being restricted to the polar regions when they did occur; compared with what *Homo sapiens* has to endure now, it was still a world of genial weather.

The change from Jurassic to Cretaceous had brought to an end the previously steady conditions, and there was an alteration of the face of the Earth by the mountain-building that took place along the Pacific border of North America. These folding movements of the crustal rock layers were the prelude to tens of millions of years of geological disturbances during which the great mountain ranges of today, such as the Andes, Alps and Himalayas, were produced, with their noticeable effect on the climate and on the animal population.

As the reptiles died out, so did the mammals increase, taking the vacated places in the various food chains. The fossil remains of the reptiles provide a much better climatic record than do those of the mammals. Since mammals are warm-blooded and automatically keep themselves at their living temperature, they can adapt to wide changes in temperature and are not restricted to close limits as were the reptiles. This we can see in mammal fossil remains, and in the world-wide distribution of animals today. Elephants are considered to be dwellers in the tropics and subtropics, but their close relatives, the mammoths, lived in arctic and sub-arctic regions. The last 70 million years, then, do not provide us with so much detailed climatic evidence from animal fossils as the earlier geological periods. However, as with all the tools that are today being applied to research into the past state of the Earth, there are compensations, in that the geological record is far less obliterated by erosion and earth movements as we approach the present day, and the more recent geological deposits are within reach of our drills and sea-bed cores.

What fossil evidence there is for the Tertiary (2–70 million years ago) suggests that the middle and northern latitudes were still experiencing warm climates, well suited to early horses, rhinoceroses and tapirs which were widely spread at least in the earlier times of this period. Towards the end of the Tertiary, some 10 to

20 million years ago, when man-like apes were appearing, the climate seems to have started its approach to what we suffer today, probably because of the changing distribution of land and oceans. The different hot and cold and wet and dry seasons of the year appear to have replaced the more equable, all-the-year-round regime of the previous several hundred million years. The philosophically inclined may feel that man came along at a time in the Earth's evolution in which he was, if not suited, at least smart enough to adapt by making overcoats and houses and refrigerators in order to twist the environment overnight, rather than undergo the slow natural process of survival of the fittest.

The final two million years of fossil history is already familiar, with the four glacial ages in the northern hemisphere and the advances and retreats of the musk-ox and reindeer which followed the movements of the ice. Some of the well-preserved animal remains found in Arctic ice must have been the result of the mammoths and their contemporaries being caught in severe blizzards at the onset of winter. Although the evidence from sea-bed cores (Chapter 6) indicates a rapid start to some ice-ages, it must be realized that 'rapid', in this geological sense, could mean a few hundred years rather than the days or hours which would be the time-scale for trapping whole animals. There is always a problem with relative periods of time when discussing Earth history. The Earth is 4,500 million years old, and the formation of new layers of rock or the erosion of old ones takes millions of years. To a geologist, it is reasonable to say that a major earthquake will occur in the San Francisco area 'any minute now', but in actual time, the 'minute' could be hundreds of years. Although, as we have seen, climate generally changes slowly, the fact that the Earth spins on its axis and orbits round the sun does provide short time-scales of days and years, so that while thinking of the slow changes that occur in the biological record, the effect of day and night and the seasons of the Earth must also be remembered.

Although the past 600 million years seem to have been characterized by a steady warm climate in which plant and animal life developed, there is some geological evidence that ice-ages did occur in the Carboniferous–Permian period 270 million years ago, and also that there were glaciations in the pre-animal days of 500–2,000 million years past. It has been suggested that the output from the sun has fluctuated owing to processes in the sun's interior which cause 'flickers' with 250- to 300-million-year periods, somewhat analogous to the sunspot occurrences with their 11-year cycle. However, if the sun was responsible for cold spells, the whole Earth would have suffered at the same time, and this does not appear to have been so for the glaciations of the Ordovician and Permian periods. It is important that account be taken of modern geological theories concerning continental drift and plate tectonics, which were introduced in Chapter 6. It was noticeable that the unsettled times of the Permian, as deduced from plant and animal fossils, were associated with changes in the shape of the Earth's surface caused by mountain-building. The detailed evidence of the positions of the various land masses, derived from measuring the direction of magnetization of old layers of rock (which took up and fossilized the north–south direction of the Earth's magnetic field when they were laid down) and examining rock type similarities in what are now different continents, shows that land movements sideways were associated with mountain-building. In fact, it is reasonable to suppose that the sideways travel of tectonic plates of the crust caused the forces that crumpled the edges of the plates to form the mountain ranges.

The evidence for colder climates in the past, apart from that provided by fauna and flora, rests on the geological traces of glaciers and ice-sheets, and such evidence would be very obvious if we could remove the ice-cover from, say, Greenland or Antarctica. Similar firm evidence of glaciation can be seen in the Alps or other ice-bound mountains. If we now accept that continents have moved around on the Earth's surface, it is probable that at some times they will have been in polar regions and subject to snow and ice, with all the associated geological marks of glaciation to be unearthed hundreds of millions of years later when the appropriate rock layers are subjected to geological examination. During most of the Permian and Triassic times (180–270 million years ago) all the land was in one giant continent, there was no Atlantic or Pacific Ocean, and South America, Antarctica, South Africa and Australia were clustered around the South Pole and were, no doubt, glaciated sufficiently to give all the evidence of an ice-age. The best geological, palaeomagnetic and plate-tectonic thought and fact has been used to show what the world looked like at various geological periods during the times since the Cambrian period. It is fascinating, from the climatological point of view, to see the large changes that have taken place in the relative distribution of land and sea, as the work of A. G. Smith and others has revealed.

At the beginning of the Ordovician, 500 million years ago, West Africa was at the South Pole, Australia was north of the equator, and a large ocean occupied most of the northern hemisphere, with the North Pole almost at its centre. Antarctica was attached to the southern part of the African continent and lay across the equator. Europe and North America were moving towards each other, and in fact joined forces 400 million years ago and welded on to northern Asia by 300 million years ago. Greenland was near the equator, at the opposite side of the world to Antarctica.

At the beginning of the Carboniferous period, most of the land was in the southern hemisphere, with the southern tip of South America at the South Pole, South America still firmly attached to

Africa and North America butting on to what is now Colombia and Venezuela, but stretching to the equator so that Hudson Bay and Greenland were right in the tropics, as was the United Kingdom, which was at the time about on the tropic of Capricorn. North-east Asia from Kamchatka to Malaysia was on its own, stretching, as it does now, from near the North Pole to the equator. In Devonian times, an ocean ring round the world spread from the equator to 30° N.

In Permian times (230–270 million years ago) all the land joined, to form one continent stretching from one pole to the other. Antarctica had taken its rightful place at the South Pole, and was joined to the South America–Africa complex, which stretched north so that north-west Africa was on the equator, as was also the southern part of North America; this was pushed against the northern coast of South America, leaving little room for the present Gulf of Mexico and the Central American states. Alaska was approaching the Arctic circle at this time, but Greenland was mostly in the subtropical zone, while the United Kingdom was just inside the tropic of Cancer. Norway and the Siberian coast were joined to northern Canada by pressing together Greenland and the north Canadian islands. This distribution of land led to a dry climate in what was then the tropical zone of a large continent. The dry area covered what is now Brazil, north-west Africa, Europe, Greenland and all of North America to the east of the Rockies. Rain and snow fell in the land which was near the poles and summer monsoons covered east Asia, which was in north latitude 10°–40°. The finer details of climate can be worked out by using the fundamentals of Earth's rotation and the ensuing air flow resulting from heating of land and sea, and observing the restrictions imposed by the land on present-day climate.

The wind circulation and the surface ocean currents must have been restricted by the land distribution, and this, no doubt, contributed to the good climate of the Mesozoic era (70–230 million years ago). The single continent of Permian times began to split up about 150 million years ago, and the Atlantic as we know it today started to form by spreading from the present-day Mid-Atlantic Ridge. There was a general drift north, especially of the African continent, and by 50 million years ago the Mediterranean and the Gulf of Mexico had formed, Alaska, northern Canada and Greenland were well into the Arctic circle, and Britain and the North Sea were some 20° south of their present position. Antarctica appeared to be finally set, even more symmetrically than in Permian times, about the South Pole. Siberia had shifted over to form, with North America and northern Greenland, a circle of land around the Arctic Sea.

This is much the pattern as we see it today, and this could have contributed to ice-ages, but there was an interesting episode in world geography about 100 million years ago before North and South America joined up, when there was a continuous sea round the Earth at about 20° N, passing through the Mediterranean and out eastwards south of the present Himalayan mountain range, and north of the triangular part of southern India. This was possible because India was still on passage from East Africa and Madagascar to its present position, and had not yet pressed up against the southern shores of the Asian continent to form the massive mountainous uplifts of northern India and Tibet.

Australia broke away from the Antarctic continent about 30 million years ago, moved back northwards, and in so doing opened up the great southern ocean, allowing the winds of the Roaring Forties to drive the west wind drift, a surface current which effectively cuts off the warm ocean currents which had formerly supplied heat from warmer waters in the north.

All things considered, then, there seem to be many pieces of circumstantial evidence which make it reasonable to suppose that, some 20 million years ago, the world would get colder. Ice in the Antarctic means a large area of white to deny access to the heat from the sun, thus adding one more straw to the other recent happenings that were described in Chapter 6 as supplementary causes of the ice-ages of the last million years. Those who do not wish to entertain the thought of another ice-age in a few thousand years' time must put their faith in 'flickering' of the sun as the cause of global heating and cooling throughout the aeons of geological time, even though there are many more probable explanations of how the present situation was arrived at.

9 'Here is the weather forecast'

The development of high-speed electronic computers has enormously increased the ability of meteorologists to predict the future state of weather systems. There are two ways of tackling the prediction problem: in one, the theoretical knowledge of how air moves under pressure and other forces is used to predict the step-by-step values of wind, temperature and moisture over finite intervals of time, starting with known initial values. This method has the advantage that it attempts to predict the evolution of real, individual weather systems, such as depressions and anticyclones, but it has the inherent disadvantage that errors introduced into the predicted state will inevitably multiply with time and effectively limit the practical usefulness of the forecasts.

The second method of prediction is statistical: relationships are sought between the values of certain parameters at one instant of time and values at an earlier time. Correlations are found between several different properties of the atmosphere and these are combined in equations which relate the value of, let us say, temperature to values of wind, temperature, pressure and so on at an earlier time. The advantage of this method is that relationships may be found spanning long periods of time, which may be very useful in prediction. On the other hand the connection between wind and temperature, or hot and cold, is not usually very precise and may have fairly large possible variations plus and minus, in spite of the mean being generally correct. Much more important is the fact that the relationships may be very complex and not stable with time, so that the relationships may have existed in the past but will not necessarily hold in the future.

Before discussing the progress which has been made with these two fundamental methods of forecasting, let us consider just how reliable climatology can be. Climatology represents an average or summary of several weather patterns in each particular season and may in itself have some usefulness in prediction.

One of the earliest and most formidable attempts to find order and rhythm in the annual course of the evolution of weather patterns was accomplished by Professor H. H. Lamb, who is the world expert in this subject. As a result of a study of fifty years of daily surface pressure maps covering the east Atlantic and western Europe he was able to distinguish periods of the year when particular weather types appeared to be dominant. It is extremely interesting to examine some of Lamb's 'singularities', as the weather types are called, in relation to the natural seasons which were discussed in Chapter 3:

23 July to 7 August: Thundery cyclonic weather over Europe and the British Isles. Cyclonic weather reaches its annual maximum over the British Isles (at about 19 times in 50 years) with frequent bursts of cool maritime air across Europe. This singularity occurs during the peak of the summer season when the hot southern Rockies ridge anchors a trough of cold air over the St Lawrence, and another trough develops downstream of this near the British Isles. Cyclonic weather is very much associated with cold troughs, and so this singularity can be explained in the context of the broad-scale hemispheric flow of summer.

5–30 September: Old wives' Indian summer anticyclones. This period is notable for anticyclones passing across the British Isles and Europe into Siberia, and marks a considerable contrast to the dominant cyclonic type earlier in summer; it occurs during the autumn season (up to 40 per cent of the years examined) when the Arctic temperature change-rate is strengthening and expanding, with deepening depressions in high latitudes. The old summer temperature picture over north-west Europe is replaced by the Arctic front near Iceland and northern Scandinavia. This radical change allows anticyclones to develop on the warm side of the Arctic front, and is directly related to the strong seasonal cooling in the Arctic.

Storminess in early January: The year's maximum frequency of westerly weather types, occurring in more than half the years studied, takes place in the winter season when the hot and cold local effects are relatively stronger, leading to the now intense temperature changes. This causes the Icelandic depressions to be most active and extensive in their influence.

February anticyclones: The combination of anticyclonic, southerly and easterly weather types affects the British Isles about one year in two, when depressions are crossing the Atlantic in low latitudes towards the Mediterranean and north-west Africa, and the Icelandic depression is less active and less influential in Europe. The anticyclones develop between two areas of atmospheric temperature changes and form extensions of the intense Siberian anticyclone at the peak of its influence.

Cold stormy periods in April and early May: Frequent northerly winds blow in the spring and early presummer seasons, when temperature changes are weakening in high latitudes. This allows a cold trough to develop in the Norwegian Sea at an accommodating distance from the semi-permanent Labrador cold trough which usually exists at that time of year.

Mid-May to Mid-June: Anticyclonic weather types across northwest Europe. The driest weather of the year is usually found during this period in many places. It has about a fifty-fifty average but of course, as all good meteorologists will say, 'We cannot be sure of it.' This period of presummer is dominated by the residual cells of Arctic air becoming slow-moving over the relatively cold ocean; anticyclones develop over north-western Europe. Also the temperature gradients around the developing hot cells further south have not yet reached their northernmost positions, so as to influence weather types in more northern areas.

18–22 June and the following fortnight: This is a time of a return of the westerlies to north-west Europe with the most regular monsoonal invasion of the Continent by cool oceanic air from the west and north-west. It is clearly associated with the development of the summer season regime with its controlling hot cells.

This brief look at certain aspects of climatology gives some idea of how regular and rhythmic the weather patterns can be in their evolution during the course of the year. In other words, climatology *can* be used as a long-range forecast and sets a standard that dynamical and statistical prediction techniques find hard to beat as yet. But this does not mean that meteorologists have given up.

Indeed, long-range forecasting of this type is still in its infancy, but progress is certainly being made.

In addition to using all the fundamental equations that have been developed by physicists since the time of Newton to account for the flow of fluids and the exchange of heat by conduction and convection it is necessary to take account of the water in the atmosphere if we are to arrive at useful predictions by theoretical calculation of what ought to happen to a mass of air on a spinning world subject to warming from the outside by the sun. Water plays a great part since it evaporates readily at the sort of temperatures that occur on Earth, and it needs heat to evaporate. That heat will be given out again when the vapour turns into rain, and so the water in the atmospheric system acts as a carrier of heat from one part of the world to another. Water has another peculiarity compared with most other materials in that it freezes at temperatures that are frequently met on Earth. Ice and snow, as we have seen, play many different roles in locking up masses of water from circulation and altering the heat received by the Earth from the sun, on account of their good reflecting power. Apart from these effects of water in the atmosphere, water also helps to add to the 'greenhouse' effect (p. 100) and probably was the original source of oxygen in the atmosphere. What would we do without water!

In order to produce practically useful theoretical forecasts the relevant interrelating equations must be solved simultaneously at a substantial number of points around the hemisphere, say every 300 miles. This requires something like a thousand points at the Earth's surface, and we also need solutions at several higher levels in the troposphere, since, as was noted in Chapter 3, the jetstreams interact with the ground-level effects. As if the task were not big enough it turns out that the time-steps used in forward working out of the equations must be closely related to the spatial distance between grid points, to prevent undesirable factors affecting the solutions. This means the time-steps are restricted to less than 15 minutes at each stage. Small wonder, therefore, that high-speed electronic computers are required if the forecast of the weather up to several days ahead is in sufficiently quick time to be practically useful.

Probably the best way of putting the facts into the machine is the numerical model in use today at the United Kingdom Meteorological Office at Bracknell in Berkshire. This model divides the troposphere into ten levels and takes into account topography, deep convection and radiation in addition to the temperature, wind and moisture. The forecast model runs to six days, using most of the northern hemisphere as its source of information. This means that no account is taken of movement of air across the equator, which is probably not important for most regions of the hemisphere during a period of six days.

How good are computer-derived forecasts up to six days? First of all, any forecast can in general only be as good as the analysis upon which it is based. It has not yet been found possible to write a programme for a computer which enables it to take into account all the information available to a competent meteorologist and produce a better analysis of the weather pattern at a certain time. For example, satellite pictures reveal a vast amount of information to meteorologists, including the precise location of depressions, anticyclones and fronts, highs and lows, distribution of cloud (moisture) and, most importantly, the areas of probable vertical motion such as convection, and upslope and downslope motion. All this information can be interpreted by the meteorologist to make his analysis as accurate as possible. One of the major tasks of weather forecasters at Bracknell is to monitor and modify the computer analysis, for no computer can better their own appraisal of all the information available.

Up to about three days the forecasts are demonstrably better than if it is assumed that the weather pattern stays fixed or if we simply use the average, which is the very least that must be attained to be practically useful. Of course any forecaster worthy of the name would not expect a computer to produce a better forecast for the first twelve hours than he could do himself, but up to forty-eight- and seventy-two-hour forecasts there is no doubt that the guidance offered by the computer tends to be of a generally high quality. This means that the movement and development of depressions and anticyclones can be predicted two to three days in advance although of course there are some failures, notably when depressions develop quickly after about twenty-four hours into the forecast period and there had been little indication at the analysis time.

These failures emphasize the vital importance of ensuring that the computer's analysis is the best that can be obtained before the forecasts are calculated.

For forecast periods of four to six days, opinions differ on how useful the computer models can be. It is generally accepted that the prediction of small-scale systems is not well handled at this range, particularly if the forecasts are verified precisely for the time to which they apply. However this is a far cry from saying that the forecasts are useless. An experienced human forecaster will often discern a trend in the evolution of the weather systems indicated by the computer. It usually happens that the computer handles best the evolution of the large, deep thermal highs and lows in the troposphere and also the position of the jetstreams.

Given the broad trend by the computer, the forecaster is able to pick out the most likely areas for possible development of small-scale depressions and their likely movement. For this reason forecasts for periods of four to six days should be regarded as forecasts of probability trends and represent the weather sequence that is most likely in the light of all information available at the time. It is certainly very useful to an oilman operating in the North Sea if he is told that there is a trend towards stormy weather in about five or six days' time after a spell of quiet anticyclonic weather; he is as interested in changing his operating plans well in advance as he is concerned that the expected weather developments may be twenty-four to thirty-six hours adrift in time of arrival.

No meteorological service in the world yet claims to have developed a dynamical model capable of producing useful forecasts for periods up to fifteen days and beyond. In any case ambition is likely to be limited to predicting the positions and intensity of the general pressure pattern in each hemisphere to which a forecaster can attach probabilities for the occurrence of storms, rainfall, heat waves, fog and ice etc. Before such long-range computations can be at all successful more understanding is required of the heat and moisture exchanges between the oceans and the atmosphere, especially in the data-sparse tropics. The new generation of meteorological satellites are expected to provide valuable data in this respect. For example, METEOSAT, which was launched in November 1977, sits over one particular spot on the equator and transmits pictures every half hour to the ground receiving stations. It is also planned to launch the latest polar orbiting TIROS N satellites during 1979, from which it is hoped to measure in considerable detail the temperature distribution with height in both the troposphere and stratosphere.

Unlike numerical models, which attempt to simulate both weather and climate, the statistical approach accepts the climate as fact and concerns itself with departures or anomalies of a given month or season from the long-term average, or climate, for that particular month or season. Most techniques used in the preparation of monthly and seasonal forecasts attempt to predict the anomalies of temperature, rainfall and pressure with respect to the average for the particular period.

In much the same way as some investors look for signs in the past of particular rises and falls of value, so in statistical forecasting we can inspect past records and look for regular trends, or for odd sequences of events, or for combinations of several peculiar circumstances which, we hope, may repeat themselves in the future. The work is not simple. A cold winter is almost equally likely in Britain to be followed by a bad or a good summer. A very dry summer *may* be the sign of a wet winter, but not necessarily so. All we can say is that over many years the temperature and the rainfall will average out. In spite of the almost even chances of the future, quite useful predictions may be made if a detailed look is made at the whole of the temperature, pressure and rainfall picture, and such a look is now practically possible given the help of the computer with its large memory and rapid information-retrieval capability. For example, the distribution of departures from the average of monthly mean surface pressure for last December is compared with the anomalies for each December contained in the information store, and the years that most closely match the current December are noted. This analogue procedure may be repeated for mean temperature and rainfall anomalies, upper air anomalies and even sea surface temperature anomalies. Other variables include the phase of the sunspot cycle, phase of the quasi-biennial oscillation (the approximate 26-month cycle in the west–east component of the wind in the tropical stratosphere) and the index of surface westerly winds around the hemisphere between chosen latitudes.

Having produced a list of years each of which is given a mark for closeness of similarity, a ranking list is drawn up from which the top group of years, say five or six, are chosen as the best analogue years. These years may not necessarily be good analogue years because there is no certainty that this December can be matched from the limited historical data. On the maxim that like follows like, the January sequels of these years are taken as the best estimate of the current January weather. Ideally, of course, the top five or six analogues would have broadly similar Januarys but in practice it is found that there are often widely differing sequels to similar monthly mean patterns. This could arise because although the monthly mean patterns are similar the sequence of weather evolution through the months could be quite different. This does not necessarily invalidate the concept of analogues because some months are made up of contrasting weather types.

Nevertheless it cannot be denied that when the analogue procedure is repeated a month later it is unusual for the same group

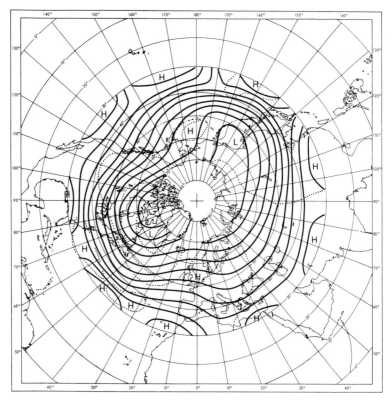

23 The long-range weather forecaster looks at today's weather and relates it to two things: the general pattern of weather behaviour over, say, the last thirty years, and the ways in which that pattern has been departed from. Here, for instance, the continuous lines represent the mean height of the 500 mb pressure surface in winter; the broken lines show the most common departures from the mean, higher than the mean (positive) or lower (negative). Thus a negative abnormality over north-east Canada means a positive one over the south-west Atlantic and Europe, and vice versa. Any particular chart can be expressed as a combination of characteristic patterns, 50 per cent of pattern one, 30 per cent of pattern two, 10 per cent of pattern three, etc. with the percentages varying from year to year

of years to be chosen as the best matches. What often happens is that one particular year of that group has, in retrospect, produced the best guidance for the current year, which implies that the real object of the analogue exercise is to pick out the winner from an apparently similar crowd!

In an attempt to solve this problem it was noted that there was often a tendency for certain analogue years to climb the ranking list over a period of successive months and then to fall away again.

This could be taken into account by noting the 'best improvers' each month and giving extra weight to those years.

In recent years a good deal of effort has gone into finding a really objective method of selecting the best of the past which is analogous to the present. It may sound a bit like pulling oneself up by one's own bootstraps, because the selection of the best comparison year of the past is made initially by the past records themselves. Let us say that we have thirty years of good surface pressure charts showing the differences from the average. A series of characteristics of the pressure patterns is tabulated and classed in order of their frequency of occurrence, number one being the pattern that occurs most times, number two the next most common and so on. An example of the use of the method is provided by the pressure charts for winter (January–February) 500 mb circulation. The pattern consists of a large anomaly centred near southern Greenland, anomalies of opposite sign south of the mean jetstream and also one centred over the English Channel. This means that when the 'average' 500 mb flow (including the jetstream) between the east Canadian low and Atlantic high is stronger than normal there is an intensification and eastward development of the low-pressure area towards southern Greenland and the high towards north-west Europe. In fact the pattern demonstrates a well-known dynamical principle in meteorology that the wavelength in the flow increases as the strength of the flow increases. Now when the coefficient of the pattern is negative the reverse situation occurs because it denotes a weakening of the flow with a smaller difference between the Labrador low and the Atlantic high. Under these circumstances the wavelengths decrease and the downstream European low is displaced westwards towards the seaboard. If all of the best characteristic patterns are compared and contrasted an interesting fact seems to emerge. Large anomalies can be seen in the same geographical regions in different patterns but what distinguishes them all is the precise combination of anomalies between one region and another. For example, the second pattern (500 mb) of autumn (September–October) implies that the Aleutian and Icelandic depressions are intensified together provided that the jetstream across central Canada is further north than normal. If the jetstream is south of normal the seventh pattern (500 mb) implies that an intensified Aleutian depression is associated with a displacement of the Icelandic depression to mid-Atlantic with an anticyclone over the Norwegian Sea. The consequences of these two patterns for the weather across the British Isles and north-west Europe are quite different!

This example is a good illustration of how a particular arrangement of anomalies in a broad sector is more important than a simple anomaly pattern in cause-and-effect relationships with other regions of the hemisphere. It may be a crumb of comfort to realize

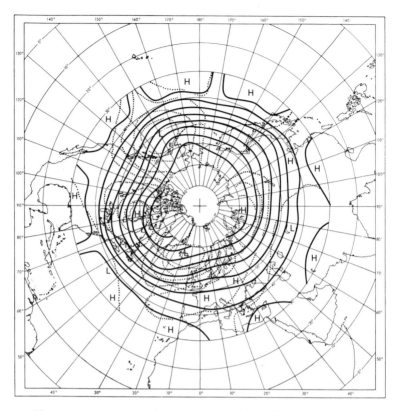

24 The most common variations are not always the most useful or informative ones. Here the broken lines show the second most common pattern of variations from the mean state in autumn

that if any nation tried to interfere with the weather of another nation by creating anomalous conditions in some remote region, it would have to create anomalies of the right sort in at least one or two other regions to ensure some degree of success.

The problem of choosing which of these analogue years is the winner is still not completely solved. One of the most infuriating and tantalizing features of the drought period over north-west Europe in 1975–76 and the subsequent very wet spell in 1976–77 was the fact that the 'good guidance' analogue years were always in the top group of years chosen, but no really objective and rational reason could be advanced for choosing the winners rather than the losers at certain critical times – for example, just prior to an exceptionally hot and dry month or just before the breakdown to cool and wet conditions. Indeed it may be that there is no unique solution and that some sequel weather types are just more likely than others. If forecasts can be expressed somehow in this manner there may be more useful information imparted than by putting all the eggs into one basket. At present the analogue methods are

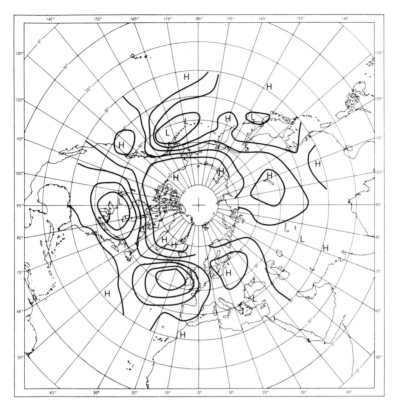

25 The seventh most common pattern of variations in autumn. All have to be looked at to see if the forecaster can find a 'fit'

generally seen as doing no more than providing clues as to the most likely sequels to the current weather situation. Because of the limiting 'goodness of fit' of any particular analogue year, subjective judgement has to be exercised on the differences between the best analogues and the current weather situation. The real trouble is that one must understand all the factors causing weather changes, and all the mathematical juggling in the world will not compensate for this lack of knowledge. Instead of simply matching coefficients in December and applying the 'like follows like' maxim for January, more precise relationships could be sought between specific patterns in December and specific patterns in January. This idea is very interesting because if the patterns can be given physical or dynamical interpretation, significant correlations would suggest that the large-scale mean weather anomalies of one month are dependent on the mean atmospheric state in the preceding month. In general it appears that there are certain regions of the hemisphere where successful predictions can be made but there are also regions of minimum predictability, one of which seems to embrace the

British Isles. This may not be surprising since the UK lies between a large ocean and an extensive continent.

There is no doubt that many meteorologists find it difficult to believe that the atmosphere can know in September what course it will take the following January. The relationships between the adjacent months or seasons can be accepted on the basis of the persistence of some physical or dynamical anomaly, for example when the areas of strong temperature change become established they can be very persistent at certain times of the year. Now if the mean atmospheric circulation for last summer happens to be identical with the average summer circulation, so that there are no anomalies, then presumably the state of the circulation during the following autumn can have a wide variety of options within the scope of what normally happens during autumn, such as rapid cooling in the Arctic with European anticyclones. But if last summer was highly anomalous in its circulation state then it may be that the number of options for the circulation state during autumn are considerably restricted. This is because anomalies in one region create anomalies in another region so that destruction of one anomaly doesn't prevent the created anomalies elsewhere from recreating the original anomaly as a kind of feedback process in some locality close to the original position.

In seeking to explain the variations from year to year in seasonal weather a good deal of effort has been devoted to studying the anomalies in sea-surface temperature (SST) and Arctic ice because of their conservative nature. Namias, the American physical oceanographer, has demonstrated numerous apparent time-lag relationships between SST anomalies in the Pacific Ocean and weather anomalies over North America. The winter of 1976–77 was one of the coldest and snowiest over the eastern USA for many years. Although not claiming to have been able to predict its severity, Namias boldly asserts that the indications of a cold winter were apparent as early as the previous November (1976) and could be indirectly linked to a persistent broad area of cold water embracing the north central Pacific together with a tongue of warm water in the extreme eastern parts of the ocean. He argues that the water temperature anomaly is transported to the atmosphere and intensifies the west to south-west upper flow over the east Pacific with consequential enhanced development of the Aleutian depression and the Rockies anticyclone.

Similar work in the United Kingdom Meteorological Office has demonstrated relationships between the north Atlantic water temperature and the weather over the British Isles. These 'teleconnections' are thought to be based upon the principle that an extra supply of heat will affect the deepening of depressions over the oceans and hence affect the surrounding circulation, but there is a strong suspicion that atmospheric circulation itself controls the heat

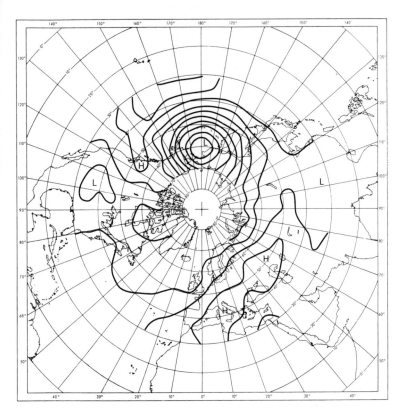

26 'Studying the patterns' can also be used to find a link between cause and effect. If we select from the ten most common abnormality patterns of winter, for instance, all those that represent strong westerly winds across the north Pacific and combine them to form a single pattern, this shows an area of positive abnormality (higher than average 500 mb level) over Europe and western Asia, causing abnormal south-westerly winds over north-west Europe. The whole world's weather patterns may be built up of linkages like this

interactions with the ocean, so that it is not certain where the argument should begin and end. This interaction between various natural processes keeps recurring in all aspects of climate and weather and it makes some research workers despair that the whole problem will ever be unravelled. However, except in cases of temperature changes of more than a few degrees centigrade covering substantial areas of the ocean it is very probable that the atmospheric circulation controls matters, particularly on the scale of a month or two. It is more likely that the SST anomalies have longer-term feedback effects, modifying the amount of sensible heat available for energy conversion in the atmospheric circulation as a whole and not confined to local areas.

One way of examining the possible effects of SST anomalies upon the atmospheric circulation is to make use of the characteristic pattern analysis. By combining all those patterns which represent the same feature in a certain region, so as to enhance that feature in that region, the resulting pattern elsewhere around the hemisphere could indicate the causal relationships. For example, certain 500 mb patterns have been combined to represent strong westerly winds across the north Pacific Ocean in each season, which could be due to SST anomalies causing stronger temperature gradients.

The winter (January–February) pattern indicates that the increased wind over the Pacific is associated with an increase of high pressure over the Mediterranean and European Russia, with increased south-westerly flow over northern Europe. Since these conditions would imply mild winters over northern Europe it could be argued that the causes of mild winters are to be found in stronger westerly winds over the north Pacific.

The amount of variability explained by this special pattern in each winter of the period 1946 to 1974 was found to exceed 25 per cent on five occasions with a peak of 45 per cent. The idea of combining the patterns in this way to highlight some possible physical or dynamical feature is potentially valuable in that it could provide some insight into our understanding of the roles of cause and effect in the complex deviations of the atmospheric circulation from year to year.

The whole problem of predicting the longer-term variations of weather is put into perspective by the drought in north-west Europe during 1975–76. It has been estimated that the sixteen-month spell of dry weather in central England was a once-in-500-years event. The associated maximum departure of surface pressure from average was located just off southern Ireland. It is calculated that one anomaly of similar magnitude occurs every year somewhere in the northern hemisphere. In other words, in terms of the normal evolution of the atmospheric circulation, there was nothing unusual about the drought. It just happened to occur where it did as a spin-off from a typical eccentricity of the circulation. It is quite a testimony to the power of the atmosphere that despite the creation of large areas of excess ice in Arctic waters east of Greenland during the late nineteen-sixties following favourable weather regimes, the atmospheric circulation reversed the processes of accumulation almost as quickly by simply changing the pattern of wind flow. The best evidence to date suggests that none of these short-period weather extremes or 'mini' climatic changes is irreversible; they are simply part of the normal expectation within our broader climatic boundaries.

One has to make the choice between believing that really long-range weather forecasts are theoretically possible, or that alternatively even the weather does not know what it is going to do next year or decade. Perhaps research effort might be spent more profitably in trying to control the weather.

10 Meddling with the weather machine

If you cannot make rain by praying for it, it is possible that you can do some good in that direction with the assistance of scientific techniques that have evolved from the study of the microprocesses that take place during cloud formation and the precipitation of water as rain, snow or hail. Silver iodide, when heated, provides a smoke of small particles which serve as nuclei on which water can form ice crystals and cause rain from a cloud that might drift away from where water was wanted. Some local success has been obtained in producing rain with particles sent up from ground heaters or by 'seeding' the clouds from the air, when frozen carbon dioxide or 'dry ice' is sometimes used instead of silver iodide. It is difficult to make sure that, after all the trouble, the rain comes down in the right place, and one can envisage occasions when neighbours will complain because an efficient operator has stolen their normal quota of rain. In general, rain-making will not have a great effect on climate because in really dry places there are no clouds to seed.

There have been proposals that cloud seeding should be used to temper the force and destructiveness of hurricanes. These giant cyclonic disturbances, about which much more has been learned since photographs of their circulation were obtained from satellites, have a beautiful heat-engine which takes up energy which the sun's rays have temporarily given to the sea water. Because of the large amount of heat given out when water vapour condenses, the full-grown hurricane can have as much power as a thousand million tons of high explosive. This potentially devastating natural weapon could theoretically be defused to some extent if the vapour could be prematurely dropped as rain before the whole whirling mass of air got out of hand. There appears to be a legal snag, however, in that if a partially destroyed hurricane proceeded to cause damage on land after being 'seeded' when it was over the ocean, a case for damages might be upheld on the grounds that the hurricane should have been left alone and harmless out at sea. The path of a hurricane is erratic, and the present state of knowledge is not good enough to forecast the track, but this is the sort of advance that research may bring, and it may make it possible to save life and property by easing the force of hurricanes, and of their smaller cousins, tornadoes, and even thunderstorms.

The existence of very large sources of almost instant energy in the form of nuclear explosions, and of more spread-out application of energy provided by large-scale civil engineering works, means that there is, for probably the first time in history, a chance for man to alter his climatic environment. Here we have two possible types of change: firstly, a purposeful attempt to alter the climate by making rain, or by stopping hurricanes; secondly, with all the energy being handled by man, there may be a chance that something untoward may occur, because of an irreversible change in environmental conditions. The early agriculturists altered their local climate by irrigation and the growing of crops after cutting down primeval forests. In the long run, with the help of peak variations in climate which appear to turn up inexorably in tens or hundreds of years, some of the arable land would turn to dust, and be drifted as moving sand dunes by the wind. A new change in the local climate could be established by planting rows of trees to break the force of the wind, and allow the soil to stabilize once more, but these local effects may not be sufficient to raise the level of agricultural production to feed a rapidly increasing human population. There is a distinct possibility, in fact, that the search for a better world food yield may lead to the sort of climatic trouble that was hinted at in Chapter 7.

A large, well-watered area of the world is occupied by tropical forests and these are being slowly cleared as civilization uses up the more temperate arable lands. At this moment, vast areas of the Amazon forests are being cut down, and gradual encroachment will occur on the African forests as the population increases. The clearing of trees has two effects on the climate of the Earth, and fortunately they work in opposite directions. Firstly, arable land in general reflects more of the sun's heat than do forests, and so large-scale change of land use could lead to a smaller total heat supply to the Earth and hence a general cooling. On the other hand, the tropical forests are great absorbers of carbon dioxide, which is needed to sustain their annual growth. Furthermore, the wood of the trees themselves is one of the large semi-fixed stores of carbon in the world, and if the forests are cleared and the timber is burnt or rotted to form carbon dioxide, there will again be a tendency to increase the carbon dioxide content of the atmosphere, and hence to heat up the Earth through the 'greenhouse effect'. The forests of the Earth have formed, with the oceans, the main stabilizers of the carbon dioxide regime, and if we wish to preserve a constant atmosphere, we should plant more trees rather than diminish the forested areas. We are in an age of great use of energy, most of which at present comes from coal and oil and is steadily adding to the carbon dioxide in the air.

The best estimates are that about 5,000 million tons of carbon (as carbon dioxide) are being added to the atmosphere every year, while an equal quantity is taken up by plant and animal life, and half as much is absorbed in the oceans. It is possible that a vast increase in minute plant life in the oceans, as probably occurred in the Carboniferous period over 300 million years ago, might provide an automatic stabilizer for the atmosphere, but the best view today is that the ocean and plant reservoirs cannot stop the drain of carbon dioxide to the atmosphere, and the Earth will inevitably warm up if we continue to increase our use of fossil fuel energy. It is probable that doubling the carbon dioxide will cause a 2°C temperature increase at the Earth's surface, although some experts think this figure is too high. This doubling will occur in about 2050, so in 70 years' time the temperature will have gone up by about 2°C. Forecasts based on population increase and the achievement of more material goods per head indicate a rapid rise of carbon dioxide during the hundred years following 2050, and some estimates give more than four times the present carbon dioxide by 2150. However, these estimates are based on a return to an increased use of coal, rather than moving forward to nuclear energy and eventually to solar power, but they do show the trend that may occur, and also the urgent need to find out more about the underlying mechanisms.

It will be recalled that a temperature increase of 2°C would bring the world back to the 'Little Optimum' of the period around AD 1100, while 8°C would probably mean the eventual removal of the polar ice-caps, a rise of 300 feet in overall ocean level, and the consequent flooding of many of the most highly populated areas of the world. While many inhabitants of the temperate regions of the northern hemisphere would probably not object to a few degrees' rise in average temperature, a rise which could well help agriculture and the world food problem, a consensus of opinion would almost certainly be against the disruption caused by the large rise in sea-level, even if this rise were very slow (because of the great heat capacity of the ocean water). It may be best to aim at a temperature rise of 3°–4°C, as occurred some four to five thousand years ago, since this caused higher evaporation in many parts of the world and thus a greater rainfall. However, even if it were politically possible to make a decision on an optimum carbon dioxide content of the atmosphere, our knowledge is not great enough at present to produce the controls. W. W. Kellogg puts the point very clearly in an article on 'Effects of Human Activities on Global Climate', in the *World Meteorological Organization Bulletin* of January 1978:

The fact is that never in the history of mankind's affairs have planners and decision-makers been given such a forewarning – with the possible exception of the Biblical story of Joseph's advice to the Pharaoh about the seven years of plenty and the seven years of famine. We have no experience of how to act given a warning of several decades.... Perhaps harbour designs and construction practices would be different if we *knew* sea-level would rise, perhaps real estate values in marginal regions would be affected if we *knew* the growing conditions would improve, perhaps new orchards would be planted with the *sure prospect* of a warmer Earth, and so forth. However, these situations are hypothetical, at least until scientists can give more assurance than they seem to feel they can give at present.

There are other things that human beings are doing today that may affect the atmosphere. There is no doubt that aircraft do lay vapour trails, but it is estimated that the Earth's surface is so much greater than the area covered by air traffic that the effect is insignificant, although a continuous cloud increase of one-fifth would be expected to give $0.5°C$ temperature increase. If aircraft were numerous enough for this coverage, there would be far more urgent problems concerned with airports and traffic control. In a similar way, the chemical constituents of jet-aircraft exhaust, and those from aerosols used on the ground, are thought to be a potential menace to the protective ozone layer in the stratosphere, but again, since these chemicals are rapidly destroyed themselves, to keep a harmful concentration in the upper atmosphere all the year round is beyond the capacity of the present chemical industry.

The dust produced by man would be expected to have some effect on the reflecting power of the atmosphere, and possibly to produce a cooling effect at the Earth's surface, but dust particles tend to settle, rather than mix uniformly with the air as carbon dioxide does. However, at the lower levels, and locally, human activities do have a great effect on climate. Smog produced by outputs from car exhausts, factories, incinerators etc. has changed many large cities such as Los Angeles from pleasant, sunny places to areas calling for strict anti-pollution legislation. The very fact of building a large conurbation has a measurable effect on the local climate, even without the smog-producing vapours. Towns produce extra heat, and the modern large buildings deviate the airflow from its normal path. The replacement of virgin soil and vegetation with roads and houses alters the heat-reflecting capacity of the surface, and also the uptake of moisture from the land, so that it comes as no surprise to hear that the rainfall and temperature patterns of an enormous city such as Houston, Texas, are changing in a lifetime. In general it appears that the centres of large urban areas are drier than the outlying districts, and that while the effects are local in character, a change in rainfall pattern could upset the original overall plans for such things as drainage of a new large city area. The importance of meteorological studies increases with the increase in world population.

It is not only the centres of large cities that tend to warm up due to human activity. Local temperature increases occur in rivers and estuaries where cooling water from power stations and factories is discharged. There may be enough heat to stimulate plant and animal growth and, if care is not taken, to use up all the oxygen to form a lifeless volume of water. On the other hand, proper use of local waste heat can provide the basis of new fish-farming industries, in much the same way as local heating of orchards has been used to keep frost away, or large flare paths were used in wartime to dispel fog. A larger problem for the future than local heating might be a contribution to a warming up of the atmosphere as a whole. At present the total heat produced by man is less than one ten-thousandth of the heat given to the Earth's surface by the sun, so man has a negligible effect on the heat balance of the Earth. However, if we take the sociologist's figures for population increase and for growth in what products each of us is going to be using, we can make a calculation that suggests that man-generated heat will cause a rise of $1°C$ by the year 2100 in temperate regions, with a figure four times this at the poles. This is not as large an increase as that calculated for the increase of carbon dioxide from burning fossil fuels, but it could be important if the nuclear age was extended to many centuries without a swing gradually to the direct use of solar energy.

There is no doubt that more food will be needed by the world in the centuries ahead, when the use of more energy per year will probably have warmed the Earth by a few degrees. This may automatically solve the problem by bringing more good wheat land under cultivation, and it is not surprising, therefore, that artificial schemes have been proposed to achieve increased fertility in various parts of the world either by warming up the land or by increasing the rainfall. One vast scheme is to dam the Ob and Yenisei rivers, which run northwards through Siberia to the Arctic Ocean. The water would be diverted to render fertile the dry steppe area north of the Caspian Sea. A secondary effect, which could be appreciated by some gourmets, would be to stop the Caspian Sea from becoming smaller and saltier, a process which is harming the production of caviar. However, the surface Arctic Ocean water might become more salty in the absence of the supply of fresh river water, and this would tend to stop surface ice formation. As noted in Chapter 6, this could have an influence on the onset of an ice-age by changing the wind circulation pattern and by providing more water for precipitation as snow to feed the Greenland glaciers.

There are local advantages, especially now that oil has been discovered on the borders of the Arctic Ocean, in freeing this area from ice so that the oilfields can be exploited and serviced. Already improved techniques of ice-breaking, using hovercraft to deform the ice ahead of strengthened ice-breakers for example, have made it possible to send large US tankers through the historic North-West Passage, and for Russian cargo ships to travel to the North Pole with a possibility of following the airlines by making a shorter polar route. It may not be a bad idea, therefore, to get rid of some of the Arctic Ocean's floating ice.

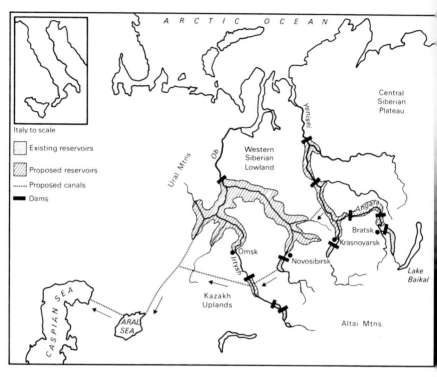

27 The plan for re-routing the rivers Ob and Yenisei to form vast reservoirs for irrigation

Another suggestion for achieving this decrease in Arctic ice is to block the fifty-mile-wide, shallow Bering Straits between Alaska and Siberia. This would stop the warm, comparatively fresh Pacific surface water from entering the Arctic Ocean, and at the same time batteries of pumps could withdraw cold water from the north and reverse the cold currents that flow into the Atlantic along the Labrador coast. This in turn would allow the warm Gulf Stream waters to move in to assist in the ice melting.

Other techniques for warming the ice-bound northern areas are to spread black material such as ash or coal-dust, or even inadvertent oil spills in order to take in more of the sun's heat, which is mostly reflected by the ice and snow. Calculations on the quantity of material required indicate that the total fossil fuel output in the form of black oil or coal dust would be needed to maintain a continuous cover over a significant area to produce anything more than a local effect. The Canadian Government Environmental Department have calculated that an Arctic oil blow-out going on for years would not have any appreciable effect on the seasonal fluctuations of the movement of the ice-sheets, although it might have serious consequences on the coastal zone amenities.

Damming of rivers and the consequent formation of large expanses of water have been considered as means of rendering fertile the desert or semi-desert areas of Africa. The Nile dam has not had a great effect on rainfall in Egypt, since evaporation from the new lake at Aswan is carried far afield before becoming rain on non-Egyptian land. However, a subtle scheme which takes advantage of the existing circulation patterns might be used to increase the rainfall in the Sahel–Sudan zone of North Africa. This is to dam the Congo at the Stanley Gorge and re-form an old lake which existed 150 million years ago, before the Congo broke through to the Atlantic Ocean. The rise in level would reverse the course of a series of rivers and would feed Lake Chad in the north. Between them the two lakes would cover a substantial one-tenth of the African continent, and this would augment the water evaporation which feeds the equatorial rain system, and thus supply an increased rainfall to places which are too sparsely watered at the present time. Another lake formation has been proposed in the Qattara depression, which, being below sea-level, could theoretically be filled by a connection to the Mediterranean. This, however, may be similar to the Aswan lake, in that the wind circulation is not suitable for enhanced local rainfall. All such grandiose schemes, which admittedly are possible with modern engineering, must be examined in the light of improved models of the weather pattern as it exists at present, coupled with the knowledge of the fundamental mechanisms that control the weather. Just as in the case of local rain-making, we must proceed with caution. In particular, in addition to basing plans on normal climate, it is necessary to look at possible extremes and see what catastrophes could be produced.

There may be something to be learned by civil engineers from the geological story of the evolution of the environment, since the effect of a large number of major changes in river patterns and local climate is recorded in the rock structures and their associated fossils. Other animals in the past have also unthinkingly affected their climate. For example, the coral-building polyp has formed lagoons which trapped sediments and provided a base for prolific marine life and eventually a source of oil discoveries today. The luxuriant plant growth in Carboniferous times may have finally sealed its own fate by lowering the carbon dioxide in the atmosphere and thus cooling the Earth's surface; possibly in the future an increase in the calcium carbonate 'bank' of the seas may tend to counteract man's use of fossil fuels. The way plants can take over man's engineering activities has been demonstrated by the enormous weed growth following the construction of the Kariba dam. The geological history book forms a continuing record of species of plant and animal that have disappeared from the face of the Earth after a period of apparently 'never having had it so good'. Perhaps all things are determined by inanimate forces, and plants and animals come and go and merely leave their prints on an inexorable system. Continental drift, added to the steady rotation of the Earth, may have determined past situations, and it may have got us boxed up with an ice-bound sea in the north and an ice-covered continent in the south, so that regular ice-ages will recur as determined by our path round the sun.

There is little doubt that we are in between ice-ages at the present time, but the ellipticity of the Earth's orbit is not so great as it has been in the past, and changes may not be quite so marked. The next really big ice advance is probably due in 5,000 to 10,000 years, and the worst that can happen in coming decades is less than the Little Ice-Age type of climate that occurred in the periods around 1650 and 1800. For those living in northern temperate countries, who are the real sufferers from ice-ages, there is such a large seasonal variation of temperature and rainfall in hot or cold periods that they will already have had frequent tastes of what is to come. Some of the forecasters of the near future, who rely on various observations mentioned in previous chapters such as sunspot activity, circulation patterns of the atmosphere, tree-rings, statistical analysis of past temperature and snow records, are more optimistic and expect the cooler period around the 1960s to be repeated about 1980, after which easier conditions will extend into the first half of the twenty-first century. In general, therefore, for the next fifty years or so, things will probably go on much as they are now, and there is definitely no immediate fear of a catastrophic ice-age descending on North America or northern Europe. It is wise, however, to think of the longer-term future and, for this reason, closer attention must be paid to all the phenomena that can affect the climate.

All the evidence points to a strong in-built stability in the climate system of the Earth, with the feedback mechanisms which have already been described. This may apply in the long run to the carbon dioxide story, but it may take thousands of years for the stabilizing forces to come into play. Meanwhile, it is possible that the ice-age trends will be swamped by the temperature rise due to the burning of fossil fuels. Although the cooling that has occurred in the latter half of the twentieth century does not support this, the worst effect of carbon dioxide is not expected until around 2050. It is possible that by that time nuclear power will have taken over with clean, quiet electric motors, rather than a return to coal. However, there is a strong enough fear of an excessive warming up of the Earth to make it imperative to put more scientific effort into studying climate and improving the new models that are technically possible with the aid of modern computers.

The research into climate mechanisms will have to be an international effort. Fortunately, the World Meteorological Organization is one of the longest-established of the United Nations agencies that arrange cooperation between different countries. The need in climate research is also interdisciplinary, and a start has already been made with extensive meteorological and oceanographical experiments to determine some of the exchanges that occur between sea and air, which clearly affect weather and climate. Further work is needed by biologists, astronomers, chemists and geologists to determine both our past climatic history and also what can be learnt from processes active now. Only in this way can good advice be given to the world about sensible and safe options for energy policies of the future.

If the research work is not started now, it is unlikely that the next generation or two will suffer, but good luck to our successors in a few hundred years' time, if we have not come up with some more definite answers to climatic problems during the next fifty years.

Acknowledgments

Aerofilms Ltd 70 bottom; Chilean Red Cross 72; Cambridge University Collection: copyright reserved 65; Librairie Ernest Flammarion 68 top; by permission of the Director, Institute of Geological Sciences 70 top, 71 top; Jean-Dominique Lajoux 71 bottom; Frank W. Lane 73 bottom; Madeleine Le Roy Ladurie 68 bottom; Museum of London 65 bottom; National Aeronautics and Space Administration/Frank W. Lane 76 bottom; Novosti Press Agency 65 top; courtesy of the *Providence Journal Bulletin* 80; Royal Danish Ministry of Foreign Affairs 69 bottom; US Air Force 78 top; US Air Force/Frank W. Lane 78 bottom; US Coast and Geodetic Survey/Frank W. Lane 79; US Department of Commerce, National Oceanic and Atmospheric Administration 75 top, 77 bottom; US Department of Commerce, Weather Bureau 77 top; US Department of Commerce, Weather Bureau/Frank W. Lane 76 top; US Forest Service/Frank W. Lane 73 top; US Geological Survey photograph no. 57–PS–66 in the National Archives, Washington DC 67; US Information Service 74; United Kingdom Meteorological Office 27–35

The diagrams were drawn by Peter Bridgewater and Stanley Paine, with the exception of that on page 83 which was drawn by Sue Ebrahim

Index

Index